P9-BIN-369

Nerve and Muscle is an introductory textbook for students taking university courses in physiology, cell biology or preclinical medicine. The first edition was highly acclaimed as a readable and concise account of how nerves and muscles work. The book begins with a discussion of the nature of nerve impulses. These electrical events can be understood in terms of the flow of ions through molecular channels in the nerve cell membrane. Then the view changes to consideration of synaptic transmission: how one nerve cell can produce changes in another nerve cell or a muscle fibre with which it makes contact. Again ionic channels are involved, but now they are opened by special chemicals released from the nerve cell terminals. The final chapters discuss the nature of muscular contraction, including especially the relations between cellular structure and contractile function. This new edition includes much new material, especially on the molecular nature and characteristics of single ionic channels, while retaining a straightforward exposition of the fundamentals of the subject.

Nerve and Muscle

Nerve and Muscle

SECOND EDITION

R. D. KEYNES

EMERITUS PROFESSOR OF PHYSIOLOGY IN THE
UNIVERSITY OF CAMBRIDGE AND
FELLOW OF CHURCHILL COLLEGE

and

D. J. AIDLEY

SENIOR LECTURER, BIOLOGICAL SCIENCES
UNIVERSITY OF EAST ANGLIA, NORWICH

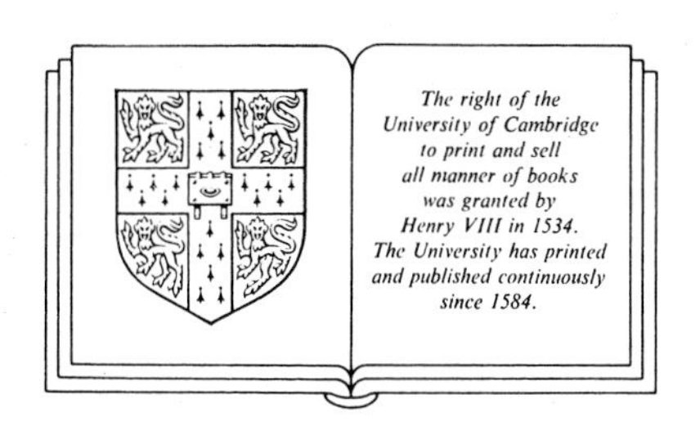

CAMBRIDGE UNIVERSITY PRESS

CAMBRIDGE
NEW YORK PORT CHESTER
MELBOURNE SYDNEY

Published by the Press Syndicate of the University of Cambridge
The Pitt Building, Trumpington Street, Cambridge CB2 1RP
40 West 20th Street, New York, NY 10011–4211, USA
10 Stamford Road, Oakleigh, Melbourne 3166, Australia

First published 1981
Reprinted 1983, 1985 (twice)
Second edition 1991

Printed in Great Britain by the University Press, Cambridge

British Library cataloguing in publication data

Keynes, R. D. (Richard Darwin)
Nerve and Muscle.—2nd. ed.
1. Vertebrates. Muscles. Nerves. Physiology
I. Title II. Aidley, D. J. (David John)
596.01852

Library of Congress cataloguing in publication data

Keynes, R. D.
Nerve and Muscle/R. D. Keynes and D. J. Aidley – 2nd edn
p.m.
Includes bibliographical references and index.
ISBN 0 521 41042 8 (hb) – ISBN 0 521 42255 8 (pb)
1. Myoneural junction. 2. Neuromuscular transmission. 3. Muscle contraction. I. Aidley. David J. II. Title.
[DNLM: 1. Muscles–physiology. 2. Nervous System–physiology. WL 102 K39n]
QP321.K44 1991
599′.01852–dc20 90-15167 CIP

ISBN 0 521 41042 8 hardback
ISBN 0 521 42255 8 paperback

Contents

Preface

In the ten years since the first edition of this book was written, important advances have been made in several of the areas with which it is concerned. These include the successful determination of the primary chemical structure of both voltage-gated and ligand-gated ion channels, the introduction of the patch-clamping technique that enables the kinetics of the opening and closing of single channels to be observed in every type of living cell and not just those long recognized as excitable, and the further unravelling of the conformational changes involved in the contraction of muscle. This second edition has accordingly been revised to bring it up-to-date, while concentrating throughout on a presentation of the basic experimental evidence in support of its conclusions. We wish again to express our gratitude to the authors and publishers who have allowed us to reproduce material that originally appeared elsewhere or has not previously been published.

R. D. Keynes
D. J. Aidley

CHAPTER ONE

Structural organization of the nervous system

NERVOUS SYSTEMS

One of the characteristics of higher animals is their possession of a more or less elaborate system for the rapid transfer of information through the body in the form of electrical signals, or nervous impulses. At the bottom of the evolutionary scale, the nervous system of some primitive invertebrates consists simply of an interconnected network of undifferentiated nerve cells. The next step in complexity is the division of the system into *sensory* nerves responsible for gathering incoming information, and *motor* nerves responsible for bringing about an appropriate response. The nerve cell bodies are grouped together to form *ganglia*. Specialized receptor organs are developed to detect every kind of change in the external and internal environment; and likewise there are various types of effector organ formed by muscles and glands, to which the outgoing instructions are channelled. In invertebrates, the ganglia which serve to link the inputs and outputs remain to some extent anatomically separate, but in vertebrates the bulk of the nerve cell bodies are collected together in the *central nervous system*. The *peripheral nervous system* thus consists of *afferent* sensory nerves conveying information to the central nervous system, and *efferent* motor nerves conveying instructions from it. Within the central nervous system, the different pathways are connected up by large numbers of *interneurons* which have an integrative function.

Certain ganglia involved in internal homeostasis remain outside the central nervous system. Together with the preganglionic nerve trunks leading to them, and the postganglionic fibres arising from them which innervate smooth muscle and gland cells in the animal's viscera and elsewhere, they constitute the *autonomic nervous system*. The preganglionic autonomic fibres leave the central nervous system in two distinct outflows. Those in the cranial and sacral nerves form the *parasympathetic* division of the autonomic system, while those coming from the thoracic and lumbar segments of the spinal cord form the *sympathetic* division.

THE ANATOMY OF A NEURON

Each neuron has a cell body in which its nucleus is located, and a number of processes or *dendrites* (Fig. 1.1). One process, usually much longer than the rest, is the *axon* or nerve fibre which carries the outgoing impulses. The incoming signals from other neurons are passed on at junctional regions known as *synapses* scattered over the cell body and dendrites, but discussion of their structure and of the special mechanisms involved in synaptic transmission will be deferred

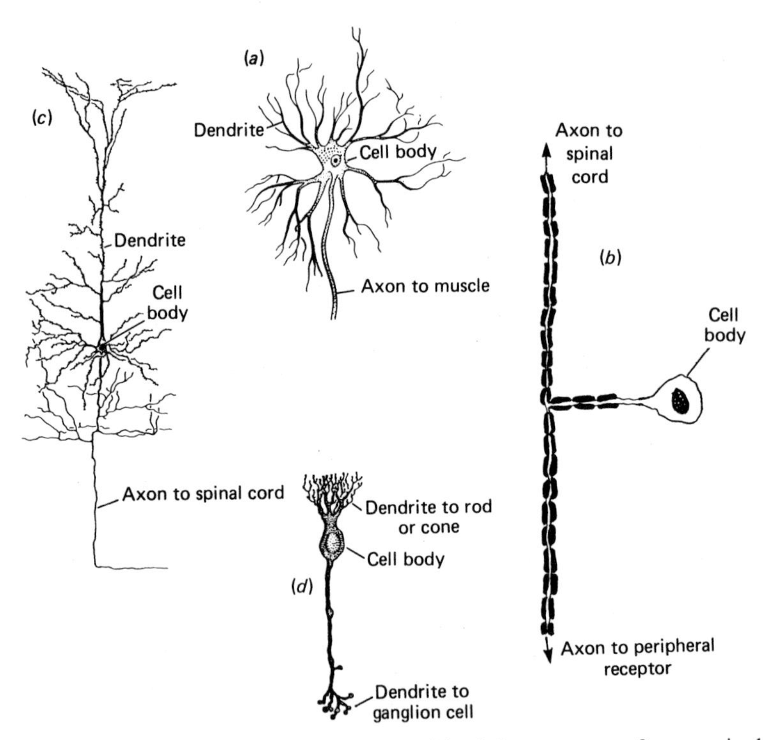

Fig. 1.1. Schematic diagrams (not to scale) of the structure of: *a*, a spinal motoneuron; *b*, a spinal sensory neuron; *c*, a pyramidal cell from the motor cortex of the brain; *d*, a bipolar neuron in the vertebrate retina.

to Chapter 7. At this stage we are concerned only with the properties of peripheral nerves, and need not concern ourselves further with the cell body, for although its intactness is essential in the long term to maintain the axon in working order, it does not actually play a direct role in the conduction of impulses. A nerve can continue to function for quite a while after being severed from its cell body, and electrophysiologists would have a hard time if this were not the case.

NON-MYELINATED NERVE FIBRES

Vertebrates have two main types of nerve fibre, the larger fast-conducting axons, 1 to 25 μm in diameter, being *myelinated*, and the small slowly conducting ones (under 1 μm) being *non-myelinated*. Most of the fibres of the autonomic system are non-myelinated, as are peripheral sensory fibres subserving sensations like pain and temperature where a rapid response is not required. Almost all invertebrates are equipped exclusively with non-myelinated fibres, but where rapid conduction is called for, their diameter may be as much as 500 or even 1000 μm. As will be seen in subsequent chapters, the giant axons of invertebrates have been extensively exploited in experiments on the mechanism of conduction of the nervous impulse. The major advances made in electrophysiology during the last fifty years have very often depended heavily on the technical possibilities opened up by the size of the squid giant axon.

All nerve fibres consist essentially of a long cylinder of cytoplasm, the *axoplasm*, surrounded by an electrically excitable *nerve membrane*. Now the electrical resistance of the axoplasm is fairly low, by virtue of the K^+ and other ions that are present in appreciable concentrations, while that of the membrane is relatively high; and the salt-containing body fluids outside the membrane are again good conductors of electricity. Nerve fibres therefore have a structure analogous to that of a shielded electric cable, with a central conducting core surrounded by insulation, outside which is another conducting layer. Many features of the behaviour of nerve fibres depend intimately on their *cable structure*.

The layer analogous with the insulation of the cable does not, however, consist solely of the high-resistance nerve membrane, owing

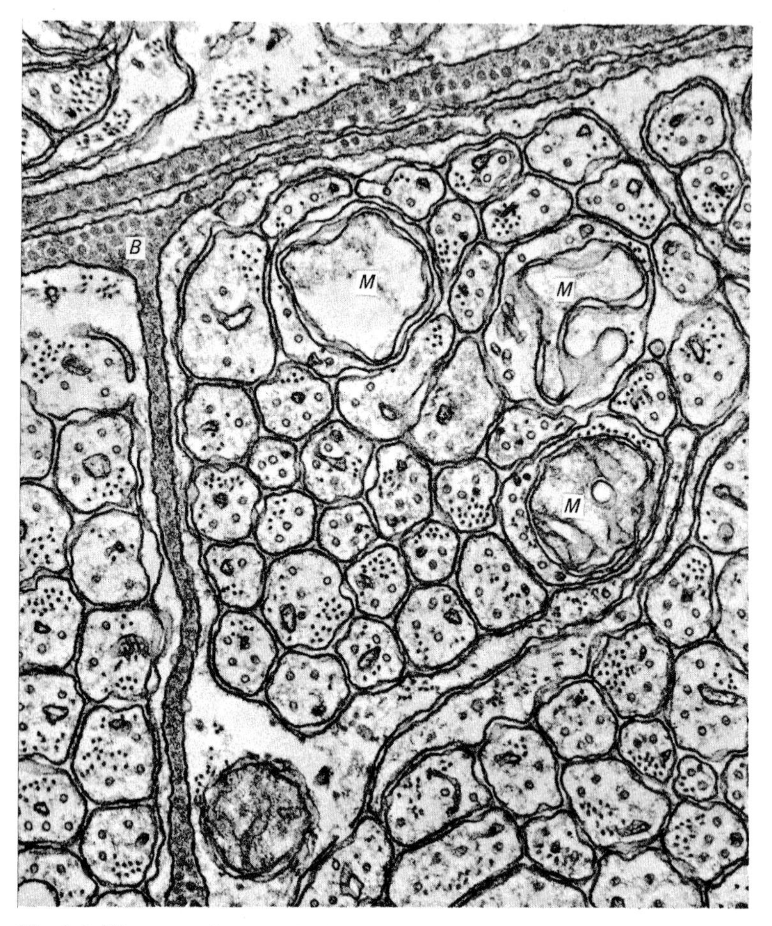

Fig. 1.2. Electron micrograph of a section through the olfactory nerve of a pike, showing a bundle of non-myelinated nerve fibres partially separated from other bundles by the basement membrane *B*. The mean diameter of the fibres is 0.2 μm, except where they are swollen by the presence of a mitochondrion (*M*). Reproduced by courtesy of Prof. E. Weibel. Magnification 48 100 ×.

to the presence of *Schwann cells*, which are wrapped around the *axis cylinder* in a manner which varies in the different types of nerve fibre. In the case of the olfactory nerve (Fig. 1.2), a single Schwann cell serves as a multi-channel supporting structure enveloping a short stretch of thirty or more tiny axons. Elsewhere, each axon may be

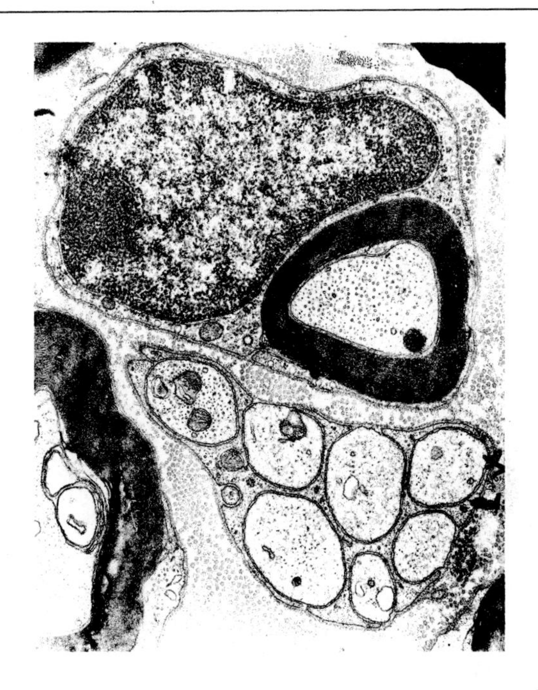

Fig. 1.3. Electron micrograph of a cross-section through a mammalian nerve showing non-myelinated fibres with their supporting Schwann cells and some small myelinated fibres. Reproduced by courtesy of Professor J. D. Robertson.

more or less closely associated with a Schwann cell of its own, some being deeply embedded within the Schwann cell, and others almost uncovered. In general, as in the example shown in Fig. 1.3, each Schwann cell supports a small group of up to half a dozen axons. In the large invertebrate axons (Fig. 1.4) the ratio is reversed, the whole surface of the axon being covered with a mosaic of many Schwann cells interdigitated with one another to form a layer several cells thick. In all non-myelinated nerves, both large and small, the axon membrane is separated from the Schwann cell membrane by a space about 10 nm wide, sometimes referred to by anatomists as the *mesaxon*. This space is in free communication with the main extracellular space of the tissue, and provides a relatively uniform

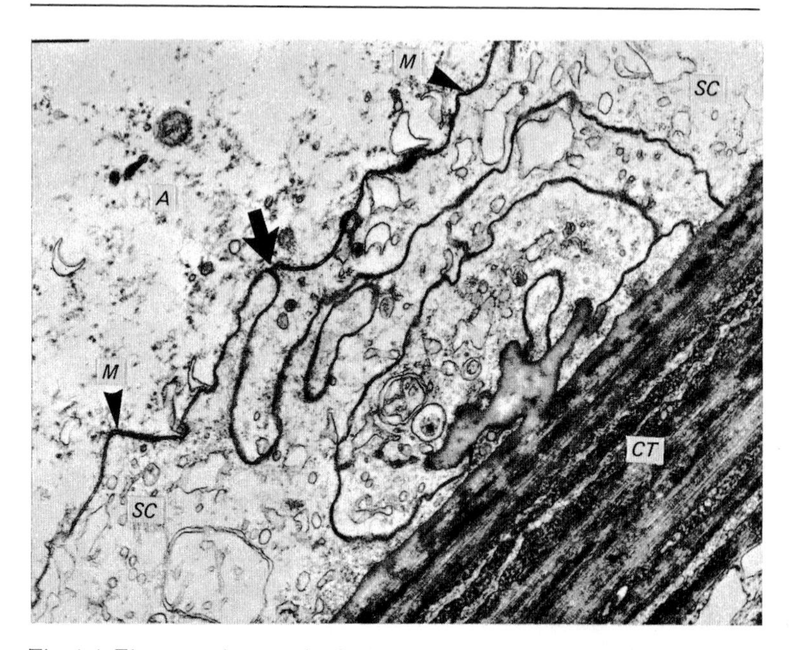

Fig. 1.4. Electron micrograph of the surface of a squid giant axon, showing the axoplasm (*A*), Schwann cell layer (*SC*), and connective tissue sheath (*CT*). Ions crossing the excitable membrane (*M*, arrowheads) must diffuse laterally to the junction between neighbouring Schwann cells marked with an arrow, and thence along the gap between the cells into the external medium. Magnification 19 800 ×. Reproduced by courtesy of Dr F. B. P. Wooding.

pathway for the electric currents which flow during the passage of an impulse. However, it is a pathway that can be quite tortuous, so that ions which move out through the axon membrane in the course of an impulse are prevented from mixing quickly with extracellular ions, and may temporarily pile up outside, thus contributing to the *after-potential* (see p. 87). Nevertheless, for the immediate purpose of describing the way in which nerve impulses are propagated, non-myelinated fibres may be regarded as having a uniformly low external electrical resistance between different points on the outside of the membrane.

MYELINATED NERVE FIBRES

In the myelinated nerve fibres of vertebrates, the excitable membrane is insulated electrically by the presence of the *myelin sheath*

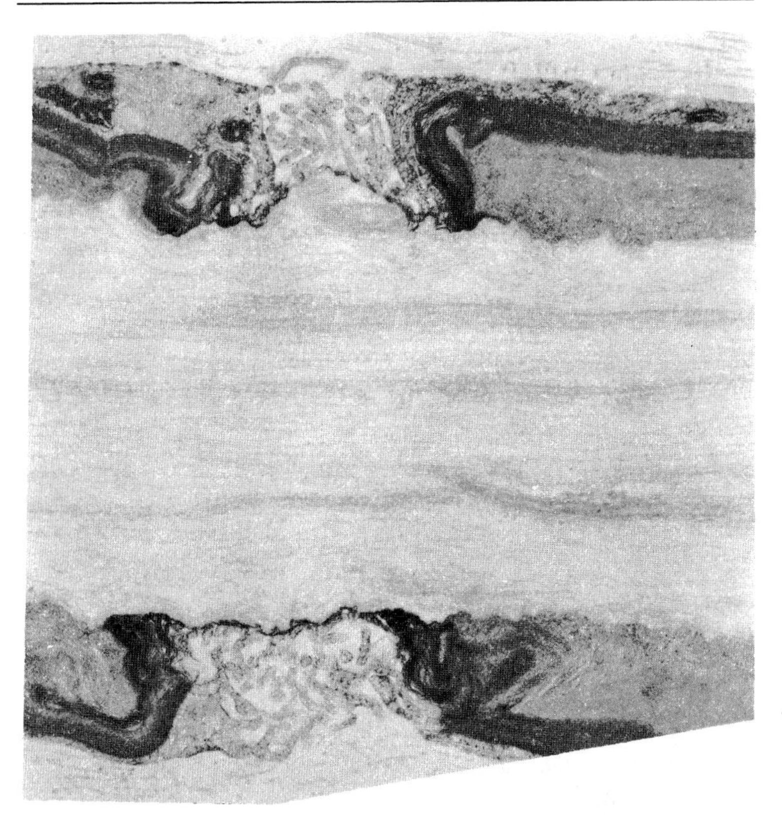

Fig. 1.5. Electron micrograph of a node of Ranvier in a single fibre dissected from a frog nerve. Reproduced by courtesy of Professor R. Stämpfli.

everywhere except at the *node of Ranvier* (Figs. 1.5, 1.6, 1.7). In the case of peripheral nerves, each stretch of myelin is laid down by a Schwann cell that repeatedly envelops the axis cylinder with many concentric layers of cell membrane (Fig. 1.7); in the central nervous system, it is the cells known as *oligodendroglia* that lay down the myelin. All cell membranes consist of a double layer of lipid molecules with which some proteins are associated (see p. 29), forming a structure that after appropriate staining appears under the electron microscope as a pair of dark lines 2.5 nm across, separated by a 2.5 nm gap. In an adult myelinated fibre, the adjacent layers of Schwann cell membrane are partly fused together at their cytoplasmic surface, and the overall repeat distance of the double membrane

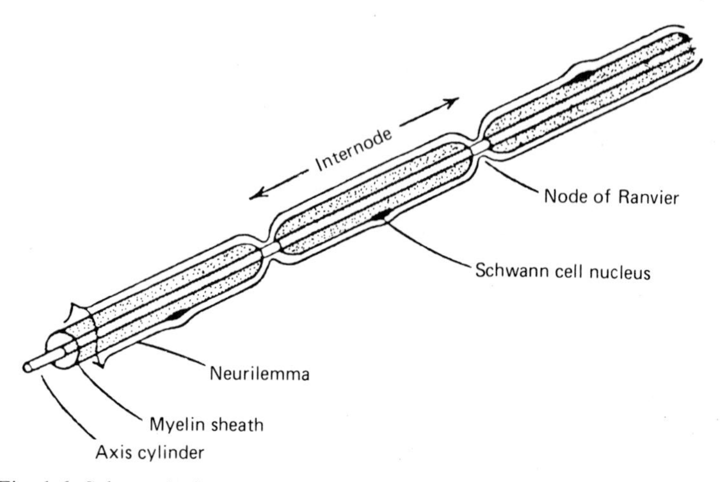

Fig. 1.6. Schematic diagram of the structure of a vertebrate myelinated nerve fibre. The distance between neighbouring nodes is actually about 40 times greater relative to the fibre diameter than is shown here.

as determined by X-ray diffraction is 17 nm. For a nerve fibre whose outside diameter is 10 μm, each stretch of myelin is about 1000 μm long and 1.3 μm thick, so that the myelin is built up of some 75 double layers of Schwann cell membrane. In larger fibres, the internodal distance, the thickness of the myelin and hence the number of layers, are all proportionately greater. Since myelin has a much higher lipid content than cytoplasm, it also has a greater refractive index, and in unstained preparations has a characteristic glistening white appearance. This accounts for the name given to the peripheral *white matter* of the spinal cord, consisting of columns of myelinated nerve fibres, as contrasted with the central core of *grey matter*, which is mainly nerve cell bodies and supporting tissue. It also accounts for the difference between the white and grey rami of the autonomic system, containing respectively small myelinated nerve fibres and non-myelinated fibres.

At the node of Ranvier, the closely packed layers of Schwann cell terminate on either side as a series of small tongues of cytoplasm (Fig. 1.7), leaving a gap about 1 μm in width where there is no obstacle between the axon membrane and the extra-cellular fluid. The external electrical resistance between neighbouring nodes of Ranvier is

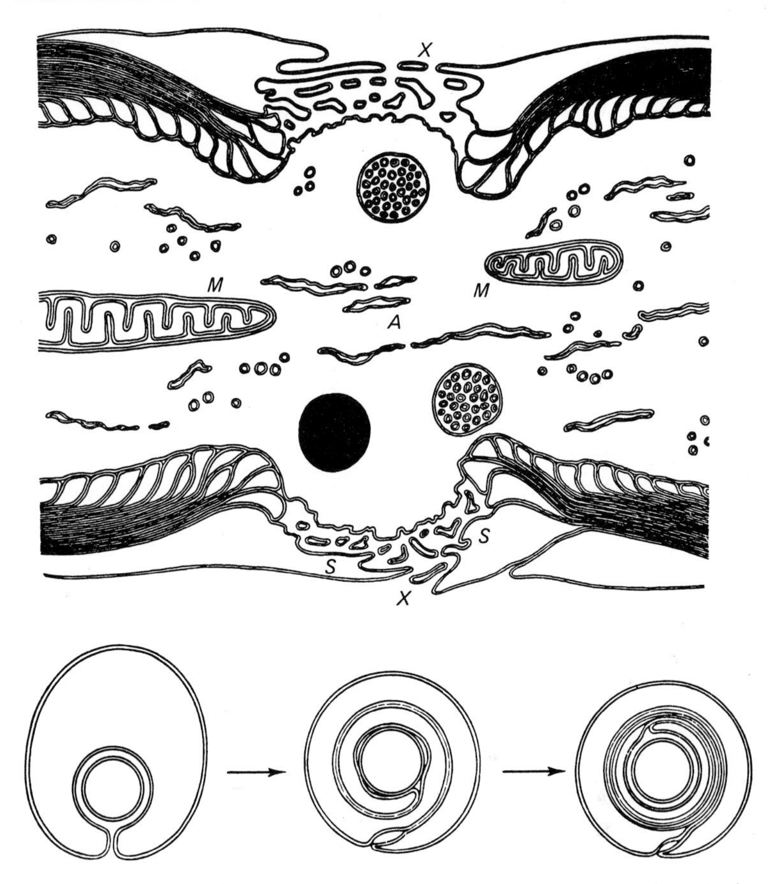

Fig. 1.7. Drawing of a node of Ranvier made from an electron micrograph. The axis cylinder *A* is continuous through the node; the axoplasm contains mitochondria (*M*) and other organelles. The myelin sheath, laid down as shown below by repeated envelopment of the axon by the Schwann cell on either side of the node, is discontinuous, leaving a narrow gap *X* where the excitable membrane is accessible to the outside. Small tongues of Schwann cell cytoplasm (*S*) project into the gap but do not close it entirely. From Robertson (1960).

therefore relatively low, whereas the resistance between any two points on the internodal stretch of membrane is high because of the insulating effect of the myelin. The difference between the nodes and internodes in accessibility to the external medium is the basis for the *saltatory* mechanism of conduction in myelinated fibres (see p. 78), which enables them to conduct impulses some 50 times faster than a non-myelinated fibre of the same overall diameter. Nerves may branch many times before terminating, and the branches always arise at nodes.

In peripheral myelinated nerves the whole axon is usually described as being covered by a thin, apparently structureless basement membrane, the *neurilemma*. The nuclei of the Schwann cells are to be found just beneath the neurilemma, at the midpoint of each internode. The fibrous connective tissue which separates individual fibres is known as the *endoneurium*. The fibres are bound together in bundles by the *perineurium*, and the several bundles which in turn form a whole nerve trunk are surrounded by the *epineurium*. The connective tissue sheaths in which the bundles of nerve fibres are wrapped also contain continuous sheets of cells which prevent extracellular ions in the spaces between the fibres from mixing freely with those outside the nerve trunk. The barrier to free diffusion offered by the sheath is probably responsible for some of the experimental discrepancies between the behaviour of fibres in an intact nerve and that of isolated single nerve fibres. The nerve fibres within the brain and spinal cord are packed together very closely, and are usually said to lack a neurilemma. The individual fibres are difficult to tease apart, and the nodes of Ranvier are less easily demonstrated than in peripheral nerves by such histological techniques as staining with silver nitrate.

CHAPTER TWO

Resting and action potentials

ELECTROPHYSIOLOGICAL RECORDING METHODS

Although the nervous impulse is accompanied by effects that can under specially favourable conditions be detected with radioactive tracers, or by optical and thermal techniques, electrical recording methods normally provide much the most sensitive and convenient approach. A brief account is therefore necessary of some of the technical problems that arise in making good measurements both of steady electrical potentials and rapidly changing ones.

In order to record the potential difference between two points, electrodes connected to a suitable amplifier and recording system must be placed at each of them. If the investigation is only concerned with action potentials, fine platinum or tungsten wires can serve as electrodes, but any bare metal surface has the disadvantage of becoming *polarized* by the passage of electric current into or out of the solution with which it is in contact. When, therefore, the magnitude of the steady potential at the electrode tip is to be measured, non-polarizable or reversible electrodes must be used, for which the unavoidable *contact potential* between the metal and the solution is both small and constant. The simplest type of reversible electrode is provided by coating a silver wire electrolytically with silver chloride, but for the most accurate measurements calomel (mercury/mercuric chloride) half-cells are best employed.

When the potential inside a cell is to be recorded, the electrode has to be very well insulated except at its tip, and so fine that it can penetrate the cell membrane with a minimum of damage and without giving rise to electrical leaks. The earliest intracellular recordings were actually made by pushing a glass capillary 50 μm in diameter longitudinally down a 500 μm squid axon through a cannula tied into the cut end (Fig. 2.1*a*), but this method cannot be applied universally. For tackling cells other than giant axons, glass microelectrodes are

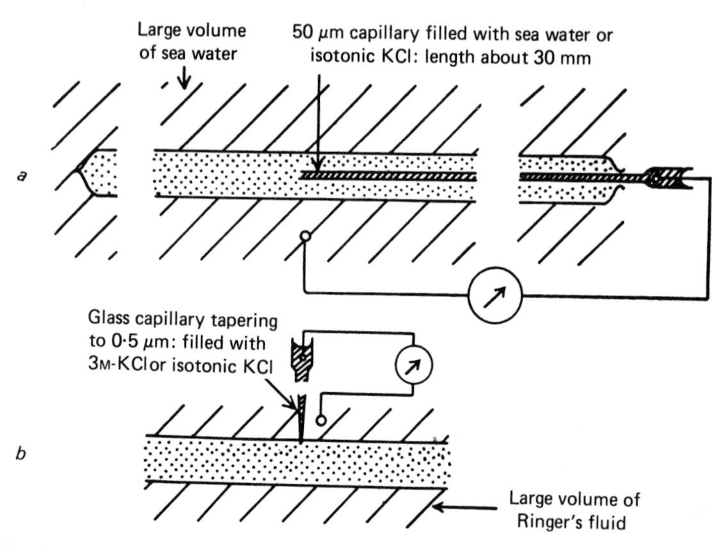

Fig. 2.1. Methods for measuring absolute values of resting potential and action potential: *a*, longitudinal insertion of 50 μm internal electrode into a squid giant axon; *b*, transverse insertion of 0.5 μm internal electrode used for recording from muscle fibres and other cells. From Hodgkin (1951).

made by taking hard glass tubing about 2 mm in diameter and drawing down a short section to produce a tapered micropipette less than 0.5 μm across at the tip (Fig. 2.1*b*). The microelectrode is then filled with 3 M-KCl, and an Ag/AgCl electrode is inserted at the wide end. With various refinements, microelectrodes of this type have been used for direct measurement of the membrane potential not only in single neurons but also in many other types of cell.

The potentials to be measured in electrophysiological experiments range from 150 mV down to a few μV, and in order to record them faithfully the frequency response of the system needs to be flat from zero to about 50 kilohertz (1 hertz = 1 cycle/s). In addition to providing the necessary degree of amplification, the amplifier must have a very high input resistance, and must generate as little electrical noise as possible in the absence of an input signal. Now that high quality solid-state operational amplifiers are readily available, there is no difficulty in meeting these requirements. The output is usually displayed on a cathode-ray oscilloscope, ideally fitted with a storage tube so that the details of the signals can be examined at leisure. To obtain a permanent record, the picture on the screen may be photographed. Direct-writing recorders yielding a continuous record on a reel of paper are convenient for some purposes, but cannot generally follow high frequencies well enough to reproduce individual action potentials with acceptable fidelity. A recent development for experiments involving close examination of the time course of the signals is to convert them into digital form, and to use an on-line computer both for storage and analysis of the data (Figs. 4.12, 4.13).

A technique that since its introduction by Hodgkin and Huxley in

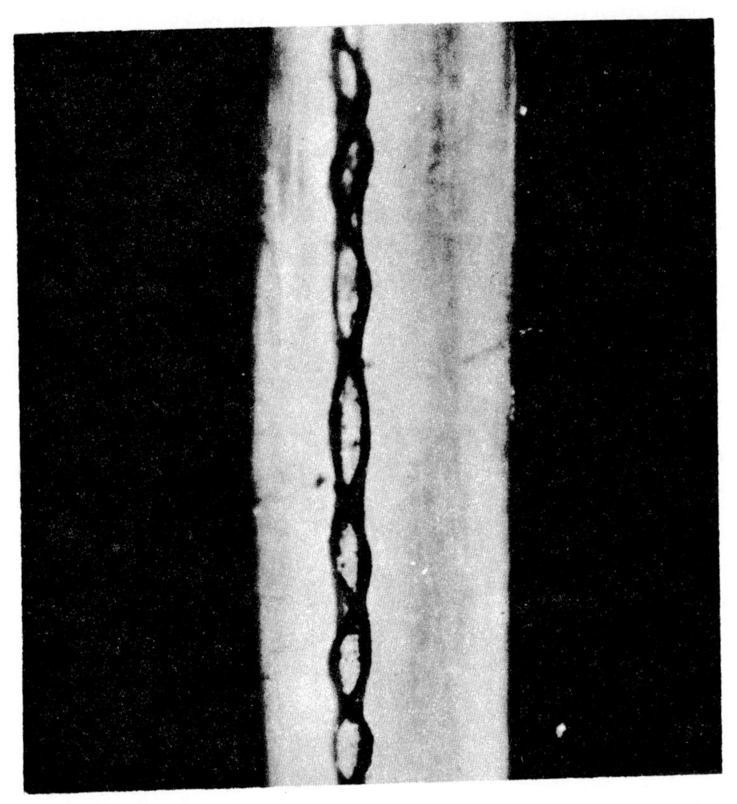

Fig. 2.2. A squid giant axon into which a double spiral electrode has been inserted, photographed under a polarizing microscope. Its diameter was 700 μm.

1949 has played an ever more important role in investigations of the mechanism of excitability in nerve and muscle is *voltage-clamping*. Its object, as explained on p. 49, is to enable the experimenter to explore the relationship between the potential difference across the membrane and its permeability to Na^+ and K^+ ions, by clamping the membrane potential at a predetermined level and then measuring the changes in membrane current resulting from imposition of a voltage step. As shown diagrammatically in Fig. 4.6, it necessitates the introduction of two electrodes into the axon, one of which monitors the membrane potential in the usual way, while the other is connected to the output of a feedback amplifier that produces just sufficient current to hold the potential at the desired value. The internal electrode system used by Hodgkin and Huxley was a double spiral of chloride-coated silver wire wound on a fine glass rod (Fig. 2.2), but others have used a glass microcapillary as the voltage electrode, to which is glued an Ag/AgCl or platinized platinum wire to carry the current. In order to voltage-clamp the node of Ranvier in a single myelinated fibre dissected from a frog nerve, an entirely different electrode system is required, but the basic principle is the same.

INTRACELLULAR RECORDING OF THE MEMBRANE POTENTIAL

When for the first time Hodgkin and Huxley measured the absolute magnitude of the electrical potential in a living cell by introducing a 50 μm capillary electrode into a squid giant axon, they found that when the tip of the electrode was far enough from the cut end it became up to 60 mV negative with respect to an electrode in the external solution. The *resting potential* across the membrane in the intact axon was thus about -60 mV, inside relative to outside. On stimulation of the axon by applying a shock at the far end, the amplitude of the *action potential* (Fig. 2.3) – or *spike*, as it is often called – was found to be over 100 mV, so that at its peak the membrane potential was reversed by at least 40mV. Typical values for isolated axons recorded with this type of electrode (Fig. 2.4*a*) would be a resting potential of -60 mV and a spike of 110 mV, that is to say an internal potential of $+50$ mV at the peak of the spike. Records made with 0.5 μm electrodes for undissected axons *in situ* in

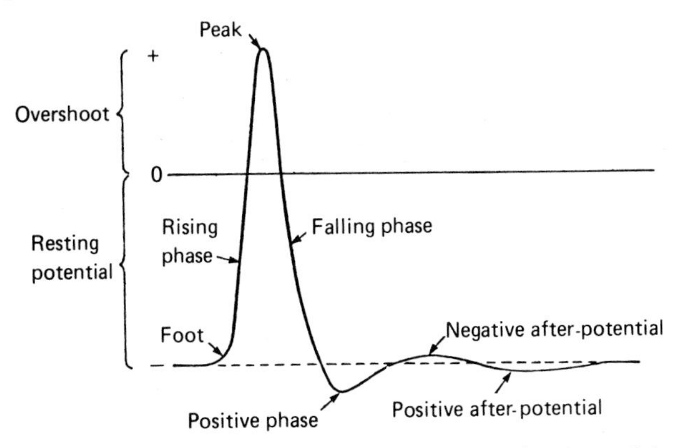

Fig. 2.3. Nomenclature of the different parts of the action potential and the after-potentials that follow it.

the squid's mantle give slightly larger potentials, and the underswing or *positive phase* at the tail of the spike is no longer seen (Fig. 2.4*b*). At 20 °C the duration of the spike is about 0.5 ms; the records in Fig. 2.4 were made at a lower temperature.

As may be seen in Fig. 2.4*c–h*, every kind of excitable tissue, from mammalian motor nerve to muscle and electric organ, gives a similar picture as far as the sizes of the resting and action potentials are concerned. The resting potential always lies between −60 and −95mV, and the potential at the peak of the spike between +20 and +50mV. However, the shapes and durations of the action potentials show considerable variation, their length ranging from 0.5ms in a mammalian myelinated fibre to 0.5s in a cardiac muscle fibre, with its characteristically prolonged plateau. But it is important to note that for a given fibre the shape and size of the action potential remain exactly the same as long as external conditions such as the temperature and the composition of the bathing solution are kept constant. As will be explained later, this is an essential consequence of the *all-or-nothing* behaviour of the propagated impulse.

EXTRACELLULAR RECORDING OF THE NERVOUS IMPULSE

There are many experimental situations where it is impracticable to use intracellular electrodes, so that the passage of impulses can only

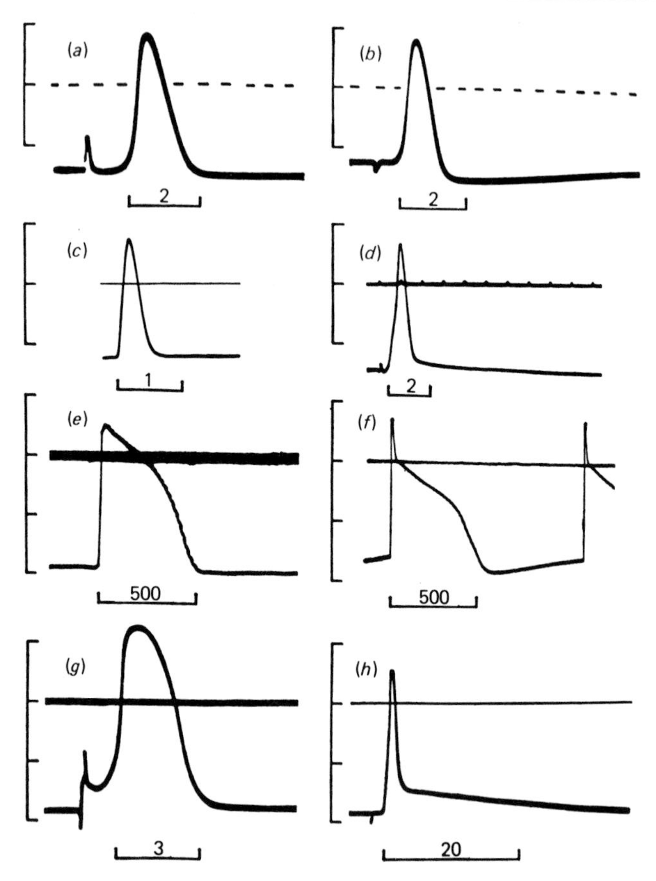

Fig. 2.4. Intracellular records of resting and action potentials. The horizontal lines (dashed in *a* and *b*) indicate zero potential; positive potential upwards. Marks on the voltage scales are 50 mV apart. The number against each time scale is its length (ms). In some cases the action potential is preceded by a stimulus artifact: *a*, squid axon *in situ* at 8.5 °C, recorded with 0.5 μm microelectrode; *b*, squid axon isolated by dissection, at 12.5 °C, recorded with 100 μm longitudinal microelectrode; *c*, myelinated fibre from dorsal root of cat; *d*, cell body of motoneuron in spinal cord of cat; *e*, muscle fibre in frog's heart; *f*, Purkinje fibre in sheep's heart; *g*, electroplate in electric organ of *Electrophorus electricus*; *h*, isolated fibre from frog's sartorious muscle. *a* and *b* recorded by A. L. Hodgkin and R. D. Keynes, from Hodgkin (1958); *c*, recorded by K. Krnjević; *d*, from Brock, Coombs and Eccles (1952); *e*, recorded by B. F. Hoffman; *f*, recorded by S. Weidmann, from Weidmann (1956); *g*, from Keynes and Martins-Ferreira (1953); *h*, from Hodgkin and Horowicz (1957).

be studied with the aid of external electrodes. It is therefore necessary to consider how the picture obtained with such electrodes is related to the potential changes at membrane level.

Since during the impulse the potential across the active membrane is reversed, making the outside negative with respect to the inside, the active region of the nerve becomes electrically negative relative to the resting region. With two electrodes placed far apart on an intact nerve as in Fig. 2.5*a*, an impulse set up by stimulation at the left-hand end first reaches R_1 and makes it temporarily negative, then traverses the stretch between R_1 and R_2, and finally arrives under R_2 where it gives rise to a mirror-image deflection on the oscilloscope. The resulting record is a *diphasic* one. If the nerve is cut or crushed under R_2, the impulse is extinguished when it reaches this point, and the record becomes *monophasic* (Fig. 2.5*b*). However, it is sometimes difficult to obtain the classical diphasic action potential of Fig. 2.5*a* because the electrodes cannot be separated by a great enough distance. In a frog nerve at room temperature, the duration of the action potential is of

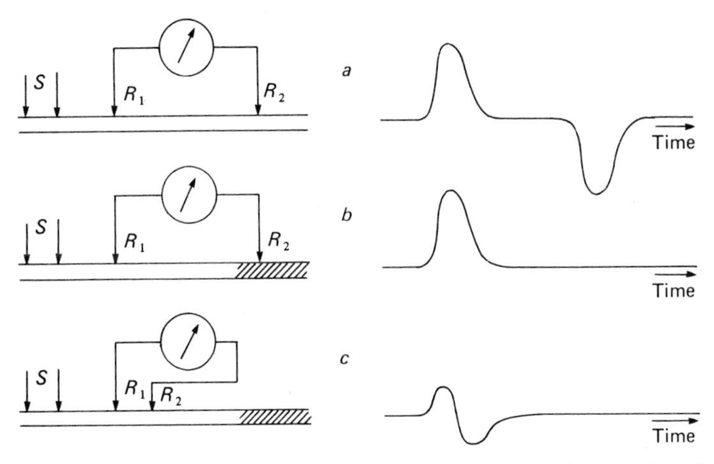

Fig. 2.5. The electrical changes accompanying the passage of a nerve impulse as seen on an oscilloscope connected to external recording electrodes R_1 and R_2. *S*, stimulating electrodes. An upward deflection is obtained when R_1 is negative relative to R_2. *a*, diphasic recording seen when R_1 and R_2 are both on the intact portion of the nerve and are separated by an appreciable distance; *b*, monophasic recording seen when the nerve is cut or crushed under R_2; *c*, diphasic recording seen with R_2 moved back on to intact nerve, much closer to R_1.

the order of 1.5 ms, and the conduction velocity is about 20 m/s. The active region therefore occupies 30 mm, and altogether some 50 mm of nerve must be dissected, requiring a rather large frog, to give room for complete separation of the upward and downward deflections. When the electrodes are closer together than the length of the active region, there is a partial overlap between the phases, and the diphasic recording has a reduced amplitude and no central flat portion (Fig. 2.5*c*).

A whole nerve trunk contains a mixture of fibres having widely different diameters, spike durations and conduction velocities, so that even a monophasic spike recording may have a complicated appearance. When a frog's sciatic nerve is stimulated strongly enough to excite all the fibres, an electrode placed near the point of stimulation will give a monophasic action potential that appears as a single wave, but a recording made at a greater distance will reveal several waves because of dispersion of the conducted spikes with distance. The three main groups of spikes are conventionally labelled A, B and C, and A may be subdivided into α, β and γ. In the experiment shown in Fig. 2.6, for which a large American bullfrog was used at room temperature, the distance from the stimulating to the recording electrode was 131 mm. If the time for the foot of the wave to reach the recording electrode is read off the logarithmic scale of Fig. 2.6*a*, it can be calculated that the rate of conduction was 41 mm/ms for α, 22 for β, 14 for γ, 4 for B and 0.7 for C. The conduction velocities in mammalian nerves are somewhat greater (100 for α, 60 for β, 40 for γ, 10 for B and 2 for C), partly because of the higher body temperature and partly because the fibres are larger.

This wide distribution of conduction velocities results from an equally wide variation in fibre diameter. A large nerve fibre conducts impulses faster than a small one. Several other characteristics of nerve fibres depend markedly on their size. Thus the smaller fibres need stronger shocks to excite them, so that the form of the volley recorded from a mixed nerve trunk is affected by the strength of the stimulus. With a weak shock, only the α wave appears; if the shock is stronger, then both α and β waves are seen, and so on. The amplitude of the voltage change picked up by an external recording electrode also varies with fibre diameter. On theoretical grounds it might be

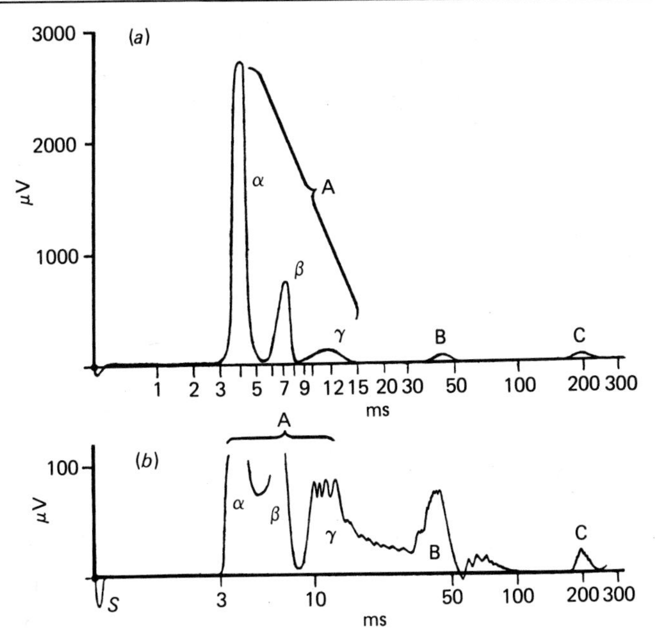

Fig. 2.6. A monophasic recording of the compound action potential of a bullfrog's peroneal nerve at a conduction distance of 13.1 cm. Time shown in milliseconds on a logarithmic scale. Amplification for *b* is ten times that for *a*. *S*, stimulus artifact at zero time. Redrawn after Erlanger and Gasser (1937).

expected to vary with the square of diameter, but Gasser's reconstructions provide some support for the view that in practice the relationship is more nearly a linear one. In either case, the consequence is that when the electrical activity in a sensory nerve is recorded *in situ*, the picture is dominated by what is happening in the largest fibres, and it is difficult to see anything at all of the action potentials in the small non-myelinated fibres.

While there is a wide range of fibre diameters in most nerve trunks, it is in most cases difficult to attribute particular functions to particular sizes of fibres. The sensory root of the spinal cord contains fibres giving A (that is α, β and γ) and C waves; the motor root yields α, γ and B waves, the latter going into the white ramus. It is generally believed that B fibres occur only in the preganglionic autonomic nerves, so that what is labelled B in Fig. 2.6 might be better classified as subdivision δ of group A. The grey ramus, containing fibres belonging to the sympathetic system, shows mainly C waves. The fastest fibres (α) are either motor fibres activating voluntary muscles or afferent fibres conveying impulses from sensory receptors in these muscles. The γ motor fibres in mammals are connected to intrafusal muscle fibres in the muscle spindles, but in amphibia they innervate 'slow' as opposed to 'twitch' muscles (see p. 123). At least some of the fibres of the non-myelinated C group convey pain impulses, but they mainly belong to postganglionic autonomic nerves. The myelinated sensory fibres in peripheral nerves have also been classified according to their diameter into group I (20 to 12μm), group II (12 to 4μm) and group III (less than 4μm). Functionally, the group I fibres are found only in nerves from muscles, subdivision IA being connected with annulo-spiral endings of muscle spindles, and the more slowly conducting IB fibres carrying impulses from Golgi tendon organs. The still slower fibres of groups II and III transmit other modes of sensation in both muscle and skin nerves.

EXCITATION

Before considering the ionic basis of the mechanism of conduction of the nervous impulse, it is best to describe some facts concerning the process of excitation, that is to say the way in which the impulse is set up in nerve and muscle fibres. This order of treatment is, historically, that in which research on the subject developed, because progress towards a proper understanding of the details of the conduction mechanism was inevitably slow before the introduction of intracellular recording techniques, whereas excitation could be investigated with comparatively simple methods such as observing whether or not a muscle was induced to twitch.

Although a nerve can be stimulated by the local application of a number of agents – for example, electric current, pressure, heat, or solutions containing substances like KCl – it is most easily and conveniently stimulated by applying electric shocks. The most effective electric current is one which flows outwards across the membrane and so *depolarizes* it, that is to say reduces the size of the resting potential. The other agents listed above also act by causing a depolarization, pressure and heat doing so by damaging the membrane. A flow of current in the appropriate direction may be brought about either by applying a negative voltage pulse to a nearby electrode, making it *cathodal*, or through *local circuit* action when an impulse set up further along the fibre reaches the stretch of membrane under consideration. It was suggested long ago that propagation of an impulse depends essentially on the flow of current in local circuits ahead of the active region which depolarizes the resting membrane, and causes it in turn to become active. The local circuit theory is illustrated in Fig. 2.7, which shows how current flowing from region *A* to region *B* in a non-myelinated fibre (upper diagram) results in movement of the active region towards the right. There are important differences that will be discussed later (see p. 78) between the current pathways in non-myelinated nerves or in muscle fibres on the one hand, and in myelinated fibres on the other (lower diagram), but the basic principle is the same in each case. The role of local circuits in the conduction of impulses has been accepted for some time, and is mentioned at this point in order to emphasize that in studying the effect of applied electric currents we are not concerned with a non-physiological and purely artificial way of setting up a nervous impulse, but are examining a process which forms an integral part of the normal mechanism of propagation.

The first concept that must be understood is that of a *threshold* stimulus. The smallest voltage which gives rise to a just perceptible

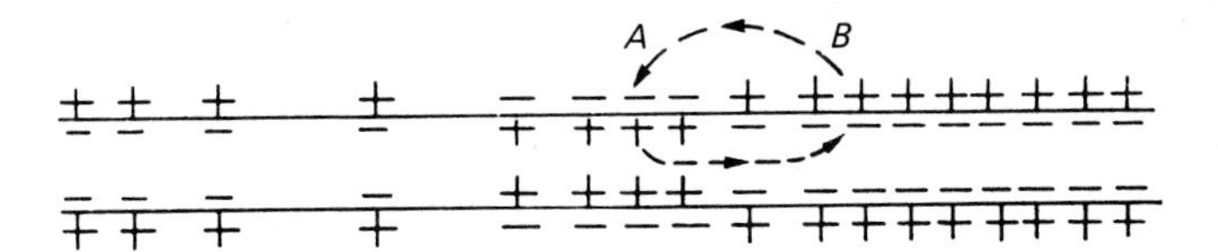

Fig. 2.7. Diagrams illustrating the local circuit theory. The upper sketch represents a non-myelinated nerve fibre, the lower sketch a myelinated fibre. From Hodgkin (1958).

muscle twitch is the minimal or threshold stimulus. It is the voltage which is just large enough to stimulate one of the nerve fibres, and hence to cause contraction of the muscle fibres to which it is connected. If the nerve consisted only of a single fibre, it would be found that a further increase in the applied voltage would not make the twitch any stronger. This is because conduction is an *all-or-nothing* phenomenon: the stimulus either (if it is subthreshold) fails to set up an impulse, or (if it is threshold or above) sets up a full-sized impulse. No response of an intermediate size can be obtained by varying the stimulus strength, though of course the response may change if certain external conditions, for example temperature or ionic environment, are altered. In a multi-fibre preparation like the sciatic nerve there are hundreds of fibres whose thresholds are spread over quite a wide range of voltages. Hence an increase in stimulus strength above that which just excites the fibre with the lowest threshold results in excitation of more and more fibres, with a corresponding increase in the size of the muscle twitch. When the point is reached where the twitch ceases to increase any further, it can be taken that all the fibres in the nerve trunk are being triggered. This requires a *maximal* stimulus. A still larger (supra-maximal) shock does not produce a larger twitch.

A good example of the threshold behaviour of a single nerve fibre is provided by the experiment shown in Fig. 2.8. Here an isolated squid giant axon was being stimulated over a length of 15 mm by applying brief shocks between a wire inserted axially into it and an external electrode, while the membrane potential was recorded internally by a second wire with a bare portion opposite the central 7 mm of the axon. The threshold for excitation was found to occur when a depolarizing shock of 11.8–12 nanocoulombs/cm^2 membrane was applied to the stimulating wire. At this shock strength, the response arose after a delay of several milliseconds during which the membrane was depolarized by about 10 mV and was in a meta-stable condition, sometimes giving a spike and sometimes reverting to its resting state without generating one. When a larger shock was applied, the waiting period was reduced, but the size of the spike did not change appreciably. The lower part of the figure shows that when the direction of the shock was reversed to give inward current which

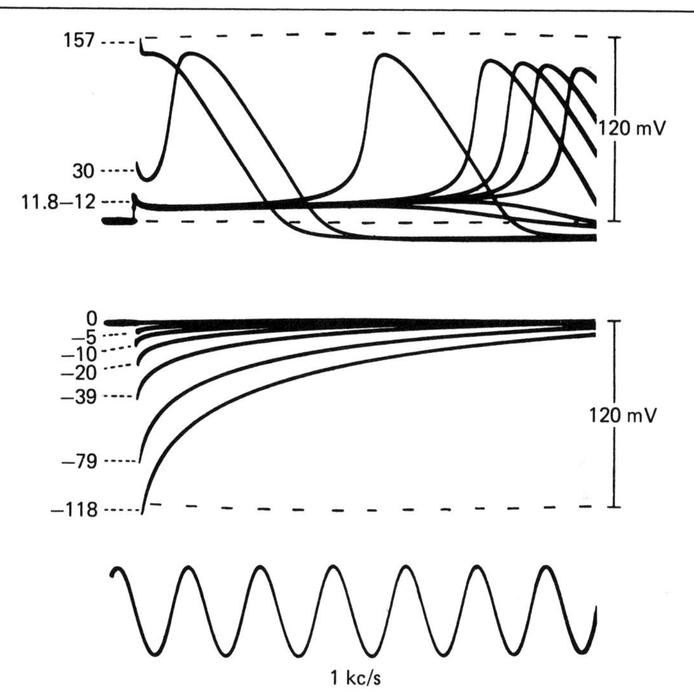

Fig. 2.8. Threshold behaviour of the membrane potential in a squid giant axon at 6° C. Shocks, whose strengths in nanocoulombs/cm^2 membrane are shown against each trace, were applied to an internal wire electrode with a bare portion 15 mm long. The internal potential was recorded between a second wire 7 mm long opposite the centre of the stimulating wire and an electrode in the sea water outside. Depolarization is shown upwards. From Hodgkin, Huxley and Katz (1952).

polarized the membrane beyond the resting level, the displacement of the potential then decayed exponentially back to the resting value. The changes in the ionic permeability of the membrane that are responsible for this behaviour are explained in Chapter 3.

An important variable in investigating the excitability of a nerve is the duration of the shock. In measurements of the threshold, it is found that for long shocks the applied current reaches an irreducible minimum known as the *rheobase*. When the duration is reduced, a stronger shock is necessary to reach the threshold, so that the *strength–duration curve* relating shock strength to shock duration takes the form shown in Fig. 2.9. The essential requirement for

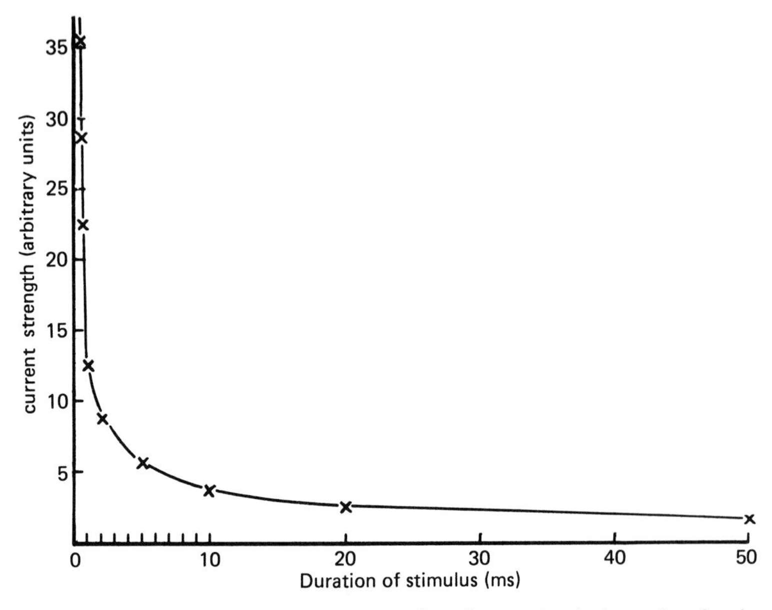

Fig. 2.9. The strength–duration curve for direct stimulation of a frog's sartorius muscle. From Rushton (1933).

eliciting the action potential is that the membrane should be depolarized to a critical level whose existence is shown clearly by Fig. 2.8. When the shock duration is reduced, more current must flow outwards if the membrane potential is to attain this critical level before the end of the shock. It follows that for short shocks a roughly constant total quantity of electricity has to be applied, and in Fig. 2.8 the shock strength was therefore expressed in nanocoulombs/cm^2 membrane.

For a short period after the passage of an impulse, the threshold for stimulation is raised, so that if a nerve is stimulated twice in quick succession, it may not respond to the second stimulus. The *absolute refractory period* is the brief interval after a successful stimulus when no second shock, however large, can elicit another spike. Its duration is roughly equal to that of the spike, which in mammalian A fibres at body temperature is of the order of 0.4 ms, or in frog nerve at 15 °C is about 2 ms. It is followed by the *relative refractory period*, during which a second response can be obtained if a strong enough shock is applied. This in turn is sometimes succeeded by a phase of super-

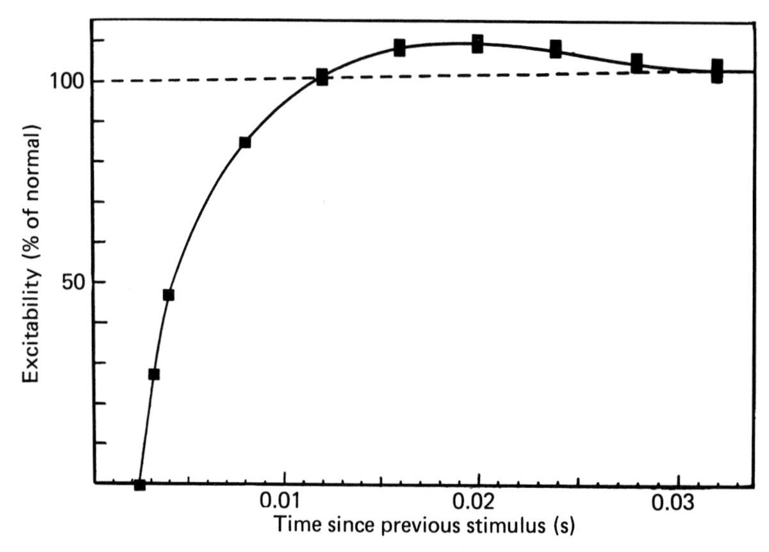

Fig. 2.10. Time course of the recovery of excitability (= 1/threshold) in a frog's sciatic nerve after passage of an impulse. The conditioning stimulus and the test stimulus were applied at electrodes 15 mm apart, so that about 0.5 ms should be subtracted from each reading to obtain the course of recovery under the test electrode. The absolute refractory period lasted 2 ms, and the relative refractory period 10 ms; they were succeeded by a supernormal period lasting 20 ms. From Adrian and Lucas (1912).

normality when the excitability may be slightly greater than normal. Fig. 2.10 illustrates the time course of the changes in excitability (= 1/threshold) in a frog sciatic nerve after the passage of an action potential.

The refractoriness of a nerve after conducting an impulse sets an upper limit to spike frequency. During the relative refractory period, both the spike size and the conduction velocity are subnormal as well as the excitability, so that two impulses traversing a long length of nerve must be separated by a minimum interval if the second one is to be full-sized. A mammalian A fibre can conduct up to 1000 impulses/s, but the spikes would be small and would decline further during sustained stimulation. In A fibres, recovery is complete after about 3 ms, so that the frequency limit for full-sized spikes is 300/s. Even this repetition rate is not often attained in the living animal, though certain sensory nerves may exceed it occasionally for short bursts of impulses.

CHAPTER THREE

The ionic permeability of the nerve membrane

STRUCTURE OF THE CELL MEMBRANE

All living cells are surrounded by a plasma membrane composed of lipids and proteins, whose main function is to control the passage of substances into and out of the cell. In general, the role of the lipids is to furnish a continuous matrix that is impermeable even to the smallest ions, in which proteins are embedded to provide selective pathways for the transport of ions and organic molecules both down and against the prevailing gradients of chemical activity. The ease with which a molecule can cross a cell membrane depends to some extent on its size, but more importantly on its charge and lipid solubility. Hence the lipid matrix can exclude completely all large water-soluble molecules and also small charged molecules and ions, but is permeable to water and small uncharged molecules like urea. The nature of the transport pathways is dependent on the specific function of the cell under consideration. In the case of nerve and muscle, the pathways that are functionally important in connection with the conduction mechanism are (1) the voltage-sensitive sodium and potassium channels peculiar to electrically excitable membranes, (2) the ligand-gated channels at synapses that transfer excitation onwards from the nerve terminal, and (3) the ubiquitous sodium pump which is responsible in all types of cell for the extrusion of sodium ions from the interior.

The essential feature of membrane lipids that enables them to provide a structure with electrically insulating properties, i.e. to act as a barrier to the free passage of ions, is their possession of hydrophilic (polar) head groups and hydrophobic (non-polar) tails. When lipids are spread on the surface of water, they form a stable monolayer in which the polar ends are in contact with the water and the non-polar hydrocarbon chains are oriented more or less at right angles to the plane of the surface. The cell membrane consists basically of two lipid monolayers arranged back-to-back with the polar head groups facing

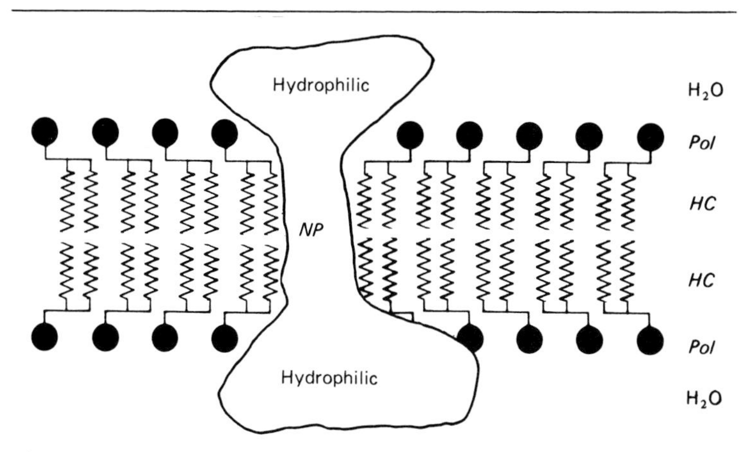

Fig. 3.1. Schematic diagram of the structure of a cell membrane. Two layers of phospholipid molecules face one another with their fatty acid chains forming a continuous hydrocarbon layer (*HC*) and their polar head groups (*Pol*) in the aqueous phase. The selective pathways for ion transport are provided by proteins extending across the membrane, which have a central hydrophobic section with non-polar side chains (*NP*), and hydrophilic portions projecting on either side.

outwards, so that the resulting sandwich interposes between the aqueous phases on either side an uninterrupted hydrocarbon phase whose thickness is roughly twice the hydrocarbon chain length (Fig. 3.1). Lipid *bilayers* of this type can readily be prepared artificially, and such so-called 'black membranes' have provided a valuable model for the study of some of the properties of real cell membranes. The chemical structure of the phospholipids of which cell membranes are mainly composed is shown in Fig. 3.2. They have a glycerol back-bone esterified to two fatty acids and phosphoric acid, forming a phosphatidic acid with which alcohols like choline or ethanolamine are combined through another ester linkage to give the neutral phospholipids lecithin and cephalin, or an amino acid like serine is linked to give negatively charged phosphatidyl serine. Another constituent of cell membranes is cholesterol, whose physical properties are similar to those of a lipid because of the $-OH$ group attached to C-3. Spin-label and deuterium nuclear magnetic resonance studies of lipid bilayers have shown that the hydrocarbon chains are packed rather loosely so that the interior

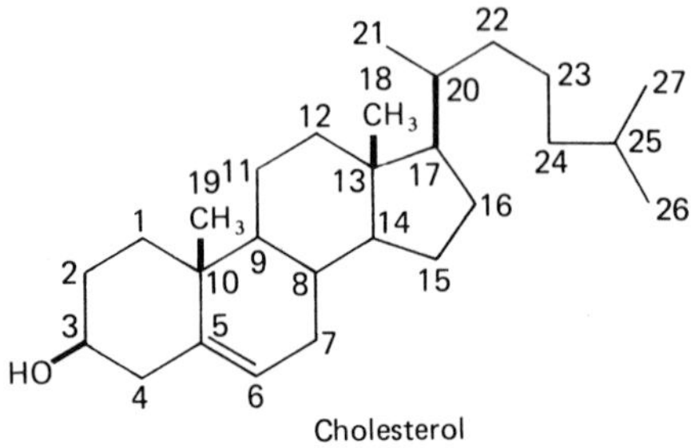

Cholesterol

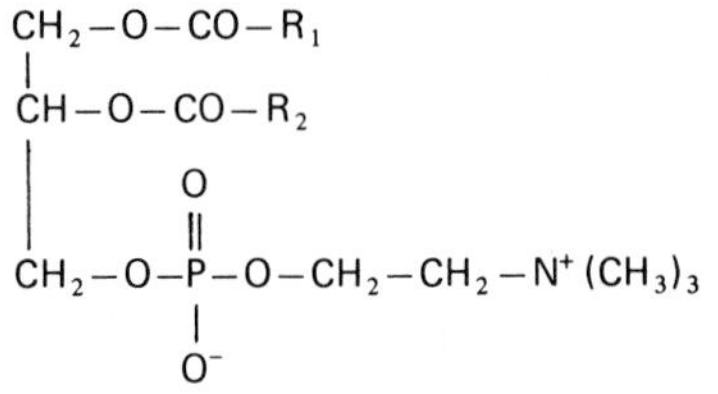

$CH_2-O-CO-R_1$
$CH-O-CO-R_2$
$CH_2-O-P(=O)(O^-)-O-CH_2-CH_2-N^+(CH_3)_3$

Phosphatidylcholine (lecithin)

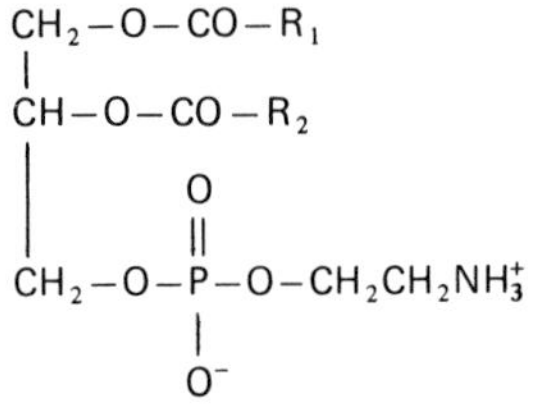

$CH_2-O-CO-R_1$
$CH-O-CO-R_2$
$CH_2-O-P(=O)(O^-)-O-CH_2CH_2NH_3^+$

Phosphatidylethanolamine (cephalin)

Fig. 3.2. The chemical structure of cholesterol and two neutral phospholipids.

of the bilayer behaves like a liquid. With a chain length of 18 carbon atoms, the effective thickness of the hydrophobic region is about 3.0 nm, which is consistent with the observed electrical capacitance of 1 $\mu F/cm^2$ membrane and a dielectric constant of 3.

Thanks to the advent of cDNA sequencing studies (see pp. 62, 99), our understanding of the organization of the protein moeity of the membrane has made rapid advances in recent years. Sections stained with permanganate or osmic acid for high resolution electron microscopy (Fig. 3.3) show the membrane in all types of

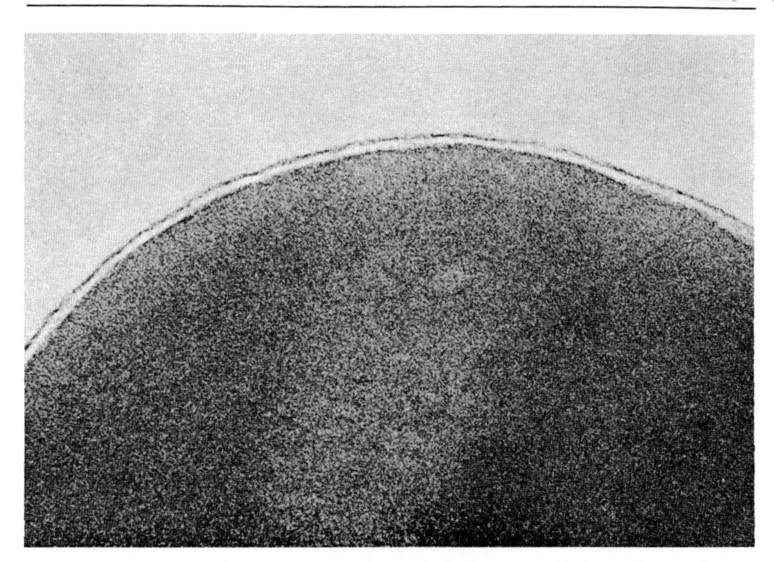

Fig. 3.3. Electron micrograph at high magnification of the cell membrane stained with osmic acid. Reproduced by courtesy of Professor J. D. Robertson.

cell to appear as two uniform lines separated by a space, the width of the whole structure being about 7.5 nm. This fits with the model proposed by Davson and Danielli, according to which the lipid bilayer is stabilized by a thin coating of protein molecules on either side, and the electron-dense stain is taken up by the polar groups of the phospholipids and of the proteins associated with them. However, an examination of freeze-fractured membranes under the electron microscope (Fig. 3.4) indicates that those proteins which traverse the bilayer to form specific ion-conducting or ion-pumping pathways are sometimes visible as globular indentations or projections. Such membrane proteins have a central non-polar section that is at home in the hydrophobic environment provided by the hydrocarbon chains of the lipids, together with polar and often glycosidic portions extending into the aqueous medium both inside and outside. Whether they are held in a fixed position in the membrane by internal fibrils, or are free to rotate and move laterally, is not always clear, but it may well be that some freedom of movement is necessary for their normal functioning.

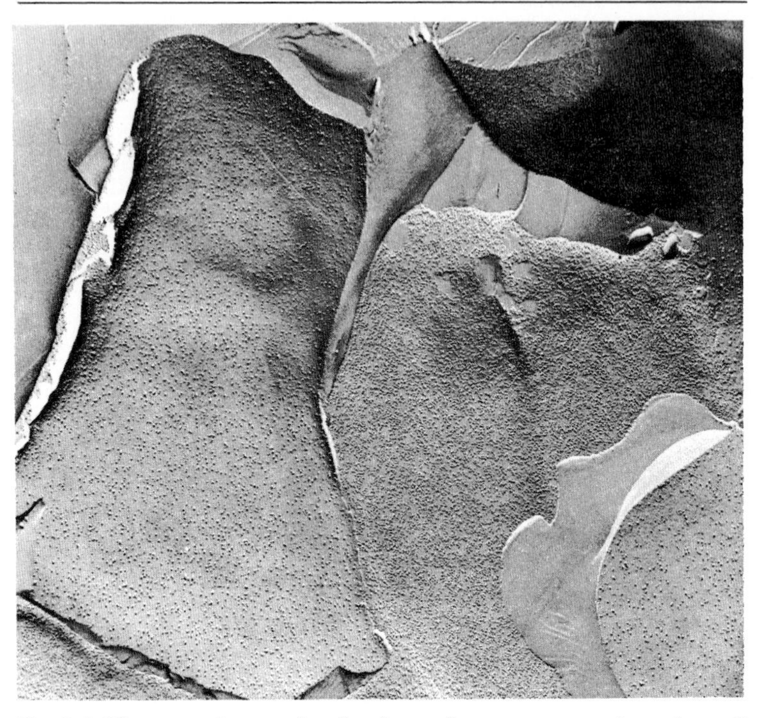

Fig. 3.4. Electron micrograph of a freeze fracture preparation of a cell membrane. The proteins appear as globular indentations. Reproduced by courtesy of Professor J. D. Robertson.

DISTRIBUTION OF IONS IN NERVE AND MUSCLE

With advent of flame photometry and other microanalytical techniques there is no difficulty in determining the quantities of ions present in a small sample of tissue. In order to arrive at the true intracellular concentrations, it is necessary to make corrections for the contents of the extracellular space, which may be done after measuring its size with the aid of a substance like inulin to which the cell membrane is impermeable. Table 3.1 gives a simplified balance sheet of the ionic concentrations in frog muscle fibres and blood plasma determined in this way.

In the case of the squid giant axon it is possible to extrude the axoplasm just as toothpaste is squeezed from a tube, and so to obtain

Table 3.1. *Ionic concentrations in frog muscle fibres and plasma*

	Concentration in fibre water (mM)	Concentration in plasma water (mM)
K^+	124	2.3
Na^+	3.6	108.8
Ca^{2+}	4.9	2.1
Mg^{2+}	14.0	1.3
Cl^-	1.5	77.9
HCO_3^-	12.4	26.6
Phosphocreatine	35.2	–
Organic anions	*c.*45	*c.*14

These figures are calculated from values given by Conway (1957). At pH 7.0, phosphocreatine carries two negative charges; the remaining deficit in intracellular anions is made up by proteins.

samples uncontaminated by extracellular ions. Table 3.2 shows the resulting ionic balance sheet.

The main features of the distribution of ions which all excitable tissues have in common are that the intracellular potassium is 20 to 50 times higher in the cytoplasm than in the blood, and that for sodium and chloride the situation is reversed. The total amount of ions is, of

Table 3.2. *Ionic concentrations in squid axoplasm and blood*

	Concentration in axoplasm (mM)	Concentration in blood (mM)
K^+	400	20
Na^+	50	440
Ca^{2+}	0.4	10
Mg^{2+}	10	54
Cl^-	123	560
Arginine phosphate	5	–
Isethionate	250	–
Other organic anions	*c.* 110	*c.*30

These values are taken from Hodgkin (1958) and Keynes (1963).

course, about four times greater in a marine invertebrate like the squid whose blood is isotonic with sea water than it is in an amphibian like the frog which lives in fresh water, but the concentration ratios are not very different. The principal anion in the external medium is chloride, but inside the cells its place is taken by a variety of non-penetrating organic anions. The problem of achieving a balance between intracellular anions and cations is most severe in marine invertebrates, and is met by the presence either of large amounts of aspartate and glutamate or, in squid, of isethionate.

THE GENESIS OF THE RESTING POTENTIAL

When a membrane selectively permeable to a given ion separates two solutions containing different concentrations of that ion, an electrical potential difference is set up across it. In order to understand how this comes about, consider a compartment within which the ionic concentrations are $[K]_i$ and $[Cl]_i$, and outside which they are $[K]_o$ and $[Cl]_o$, bounded by a membrane that can discriminate perfectly between K^+ and Cl^- ions, allowing K^+ to pass freely but being totally impermeable to Cl^-. If $[K]_i$ is greater than $[K]_o$ there will initially be a net outward movement of potassium down the concentration gradient, but each K^+ ion escaping from the compartment unaccompanied by a Cl^- ion will tend to make the outside of the membrane electrically positive. The direction of the electric field set up by this separation of charge will be such as to assist the entry of K^+ ions into the compartment and hinder their exit. A state of equilibrium will quickly be reached in which the opposed influences of the concentration and electrical gradients on the ionic movements will exactly balance one another, and although there will be a continuous flux of ions crossing the membrane in each direction, there will be no further net movement.

The argument may be placed on a quantitative basis by equating the chemical work involved in the transfer of potassium from one concentration to the other with the electrical work involved in the transfer against the potential gradient. In order to move 1 gram-equivalent of K^+ from inside to outside, the chemical work that has to be done is $RT \log_e \frac{[K]_o}{[K]_i}$. The corresponding electrical work is $-EF$, where E is the membrane potential, inside relative to outside, and F is the charge carried by 1 gram-equivalent of ions. At equilibrium, no net work is done, and the sum of the two is zero, whence

$$E = \frac{RT}{F} \log_e \frac{[\mathrm{K}]_o}{[\mathrm{K}]_i} \tag{3.1}$$

This relationship was first derived by the German physical chemist Nernst in the nineteenth century, and the equilibrium potential E_K for a membrane permeable exclusively to K^+ ions is known as the *Nernst potential* for potassium. The values of R and F are such that at room temperature the potential is given by

$$E_K = 25 \log_e \frac{[\mathrm{K}]_o}{[\mathrm{K}]_i} \mathrm{mV} = 58 \log_{10} \frac{[\mathrm{K}]_o}{[\mathrm{K}]_i} \mathrm{mV} \tag{3.2}$$

An *e*-fold change in concentration ratio therefore corresponds to a 25 mV change in potential, or a tenfold change to 58 mV.

On examining the applicability of the Nernst relation to the situation in nerve and muscle, it is found (Fig. 3.5) that it is well obeyed at high external potassium concentrations, but that for small values of $[\mathrm{K}]_o$ the potential alters less steeply than eqn 3.2 predicts. It is evident that the membrane does not in fact maintain a perfect selectivity for potassium over the whole concentration range, and that the effect of the other ions which are present must be considered. In order to derive a theoretical expression relating the membrane potential to the permeabilities and concentrations of all the ions in the system, whether positively or negatively charged, some assumption has to be made as to the manner in which the electric field varies within the membrane. Such an expression was first produced by Goldman, who showed that if the field was the same at all points across the membrane, the potential was given by

$$E = \frac{RT}{F} \log_e \frac{P_K[\mathrm{K}]_o + P_{Na}[\mathrm{Na}]_o + P_{Cl}[\mathrm{Cl}]_i}{P_K[\mathrm{K}]_i + P_{Na}[\mathrm{Na}]_i + P_{Cl}[\mathrm{Cl}]_o} \tag{3.3}$$

where the Ps are permeability coefficients for the various ions, and the suffixes o and i indicate external and internal concentrations respectively. Although the *constant field equation* has been shown in

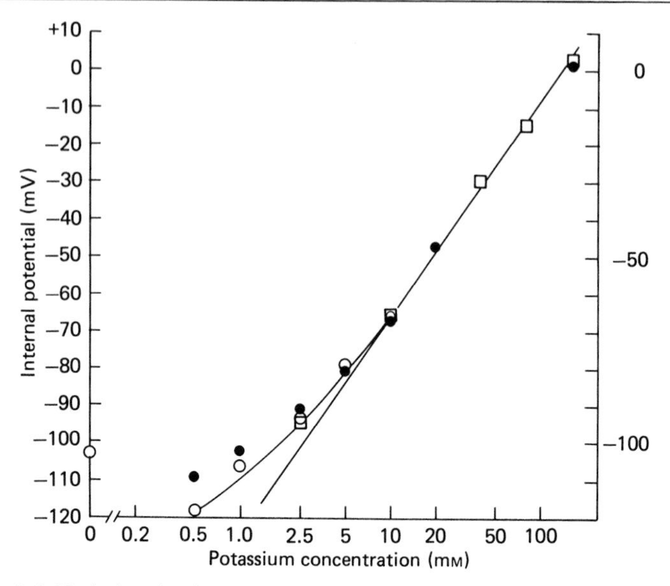

Fig. 3.5. Variation in the resting potential of frog muscle fibres with the external potassium concentration $[K]_o$. The measurements were made in a chloride-free sulphate-Ringer's solution containing 8 mM-$CaSO_4$. Square symbols are potentials measured after equilibrating for 10 to 60 min; circles are potentials measured 20 to 60 s after a sudden change in concentration, filled ones after increase in $[K]_o$, open ones after decrease in $[K]_o$. For large $[K]_o$s the measured potentials agree well with the Nernst equation, $V = 58 \log \frac{[K]_o}{140}$, taking $[K]_i$ as 140 mM. The deviation at low $[K]_o$ can partly be explained by taking $P_{Na}/P_K = 0.01$, so that $V = 58 \log \frac{[K]_o - 0.01[Na]_o}{140}$. From Hodgkin and Horowicz (1959).

practice to fit rather well with experimental observation over a wide range of conditions, it does not follow that the field is indeed truly constant. Eqn 3.3 is, nevertheless, empirically very valuable for describing the behaviour of a membrane permeable to more than one species of ion. Thus in the experiment of Fig. 3.5, the deviation of the measured potential from a line with a slope of 58 mV can be accounted for nicely by taking P_{Na} to be one hundred times smaller than P_K.

THE DONNAN EQUILIBRIUM SYSTEM IN MUSCLE

An important advance towards an understanding of the inequalities in ionic distribution observed in muscle was made in 1941 when Boyle and Conway pointed out that the type of equilibrium for diffusible and nondiffusible ions characterized by Donnan might apply. In a Donnan system consisting of two compartments separated by a membrane, the concentration ratios for any diffusible ions must be equal at equilibrium, since the same membrane potential is common to all of them. Boyle and Conway (1941) showed experimentally that in frog sartorius muscle the relationship

$$\frac{[\mathrm{K}]_o}{[\mathrm{K}]_i} = \frac{[\mathrm{Cl}]_i}{[\mathrm{Cl}]_o} \tag{3.4}$$

was duly obeyed, the ratio for K^+ ions being the inverse of that for Cl^- ions because of their opposite charges. Subsequent observations by Hodgkin and Horowicz on the effect of sudden changes in $[\mathrm{Cl}]_o$ on the membrane potential of frog muscle fibres have borne out their conclusions in every respect. A further requirement for the operation of a Donnan equilibrium is the presence of sufficient nondiffusible ions to achieve both an electrical balance between the anions and cations in each compartment, and an osmotic balance between the total solutes on the two sides. This can be met if the bulk of the anions inside the cell are unable to move outwards, and the principal cation in the external medium is unable to penetrate the cell. Boyle and Conway's proposition that K^+ and Cl^- could be regarded as diffusible ions and Na^+ as nondiffusible therefore went an appreciable way towards explaining the observed facts.

THE ACTIVE TRANSPORT OF IONS

The Donnan equilibrium hypothesis required that the muscle membrane should be completely impermeable to sodium. When the radioactive isotope ^{24}Na became available, this was soon found not to be so, for about half of the intracellular sodium in the fibres of a frog's sartorius muscle turned out to be exchanged with the sodium in the external medium in the course of one hour. Moreover, experiments on giant axons from squid and cuttlefish showed that after dissection there was a steady gain of sodium and loss of potassium that if not counteracted would eventually have led to an equalization of the sodium and potassium contents of the axoplasm. It became clear that in actuality the resting cell membrane does have a finite permeability of Na^+ ions, but that the inward leakage of sodium is offset by the operation of a *sodium pump* which extrudes sodium at a rate which ensures that in the living animal $[Na]_i$ is kept roughly constant. As far as sodium and potassium are concerned, the resulting situation should be described as a steady state rather than an equilibrium, though for experiments like those carried out by Boyle and Conway the effect is the same. Since the expulsion of Na^+ ions from the cell takes place against both an electrical gradient and a concentration gradient, it involves the performance of electrochemical work and requires a supply of energy from cell metabolism. The process is therefore termed *active transport.*

Giant axons have provided particularly favourable experimental material for radioactive tracer studies on the mechanism of the sodium pump. Fig. 3.6 shows the results of an experiment in which the sodium efflux from a *Sepia* axon loaded with ^{24}Na and bathed in a

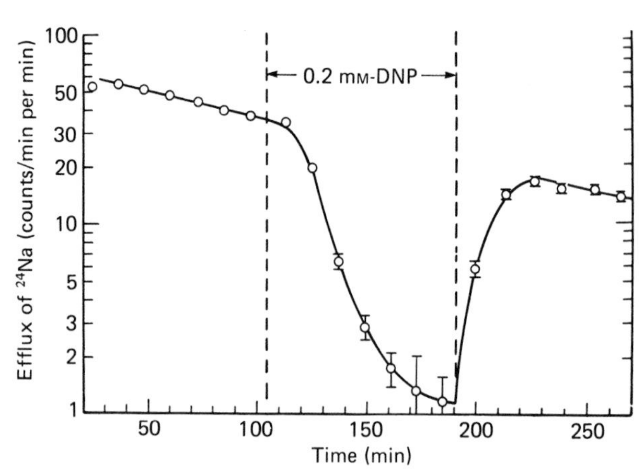

Fig. 3.6. The effect on sodium efflux of blocking metabolism in a *Sepia* (cuttlefish) axon with dinitrophenol. At the beginning and end of the experiment the axon was in unpoisoned artificial sea water. Temperature 18°C. From Hodgkin and Keynes (1955).

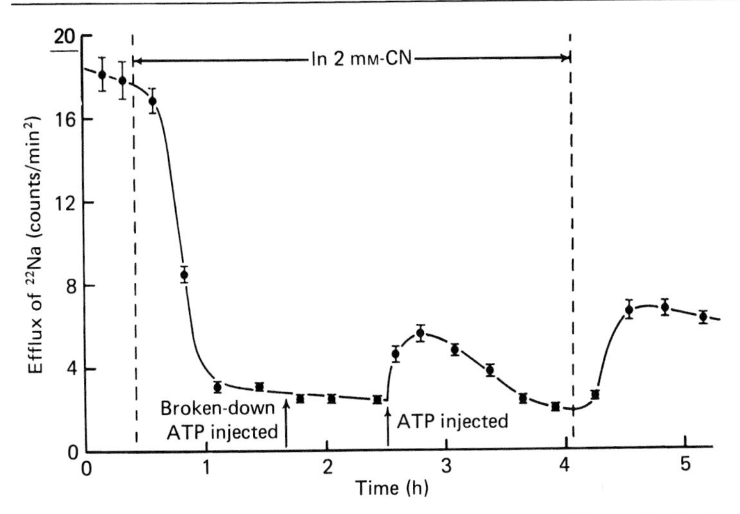

Fig. 3.7. The rate of loss of radioactivity from a 780 μm squid axon loaded by micro-injection with 6700 counts/min of ^{22}Na, distributed over 12 mm. 32 nanomoles of ATP were injected over the same 12 mm. Temperature 19 °C. From Caldwell and Keynes (1957).

non-radioactive medium was measured by counting samples of the bathing solution collected at 10-min intervals. The resting efflux was found to be roughly constant when calculated in moles of sodium per unit area of membrane per unit time, its average value at room temperature being around 40 pmole/cm^2 s. The linear decline seen in Figs. 3.4 and 3.7 when the actual counts are plotted arises from the gradual dilution of the internal radioactivity by inactive sodium entering the axon as the experiment proceeds. When, however, the metabolic inhibitor 2:4-dinitrophenol (DNP) was added to the external medium, the counting rate fell quickly to about one thirtieth of its previous level. The effect was reversible, and on washing away the DNP the efflux soon recovered. Axons treated with cyanide or azide behaved in a similar fashion. Since all these inhibitors are known to act by blocking the production of the energy-rich compound adenosine triphosphate (ATP) by oxidative phosphorylation in the mitochondria, the implication was that the sodium pump was driven by energy derived from the terminal phosphate bond of ATP.

The role of ATP as the immediate source of energy for sodium extrusion was further examined by testing its ability to restore the sodium efflux when injected into cyanide-poisoned axons. Fig. 3.7 shows that ATP injection did bring about some degree of recovery of the efflux, but it turned out that a complete recovery was only obtained if the ratio of [ATP] to [ADP] in the axoplasm was made reasonably large. This could be achieved by the injection of arginine phosphate, which serves as a reservoir for high energy phosphate in the tissues of invertebrates through the reaction

$$\text{ADP} + \text{arginine phosphate} = \text{ATP} + \text{arginine.}$$

An important characteristic of the sodium pump is its dependence for normal working on the presence of potassium in the external medium. Thus when at the beginning of the experiment illustrated in Fig. 3.8 $[K]_o$ was reduced to zero, the unpoisoned sodium efflux fell to about quarter of its normal size. After the cyanide had taken effect, a large amount of arginine phosphate was injected into the axon. This duly brought the efflux back to normal,

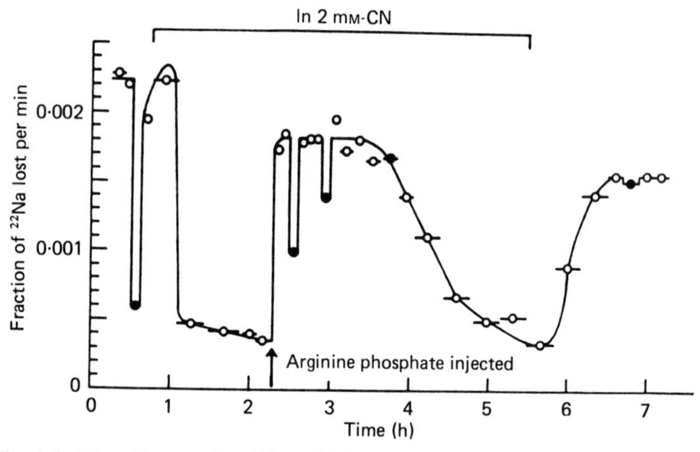

Fig. 3.8. The effect on the efflux of labelled sodium from a squid giant axon of first blocking metabolism with cyanide and then injecting a large quantity of arginine phosphate. Open circles show efflux with $[K]_o = 10$ mM; filled circles show efflux into a potassium-free solution. Immediately after the injection the mean internal concentration of arginine phosphate was 33 mM. Temperature 18 °C. From Caldwell, Hodgkin, Keynes and Shaw (1960).

but only during the first hour was it sensitive to the removal of potassium. In a similar way, the efflux that reappeared on washing away the cyanide only regained its potassium sensitivity when sufficient time had been allowed for the ATP/ADP ratio to return to normal. The requirement of the sodium pump for external potassium suggested that there might be an obligatory coupling between the extrusion of sodium and an uptake of potassium. Parallel measurements of the efflux of ^{24}Na and the influx of ^{42}K have shown that this is indeed the case, and that in many tissues the coupling ratio is normally 3:2, i.e. for every 3 Na^+ ions that leave the cell, 2 K^+ ions are taken up. If the coupling ratio were exactly 1:1, the sodium pump would be electrically neutral in the sense that it would bring about no net transfer of charge across the membrane. A coupling ratio greater than unity implies that the sodium pump is electrogenic, and causes a separation of charge which tends to hyperpolarize the membrane. The after-potential that succeeds the impulse in small non-myelinated nerve fibres is thought to arise in this way from an acceleration of the sodium pump, and the pumping of ions in many other situations has now been shown to be electrogenic to some degree.

The sodium pump occurs universally in the cells of higher animals, and can be identified with the enzyme system Na, K-ATPase first extracted from crab nerve by Skou. Research on the chemistry of Na, K-ATPase has depended heavily on the exploitation of the inhibitory action of glycosides like ouabain and digoxin, which in micromolar concentrations block both the active fluxes of sodium and potassium in intact tissues, and the splitting of ATP by purified enzyme preparations. By measuring the binding of ouabain labelled with tritium, it is possible to estimate the number of sodium pumping sites in unit area of membrane, assuming that each site binds one molecule of ouabain. In the squid giant axon there are several thousand sites per μm^2 of membrane, while in the smallest non-myelinated fibres the density of sites is about a tenth as great.

Although the extrusion of Na^+ and intake of K^+ ions by the sodium pump is quickly halted by ouabain or any metabolic inhibitor which deprives the pump of its supply of ATP, neither treatment has any immediate effect on electrical excitability. Fig. 3.9 shows the

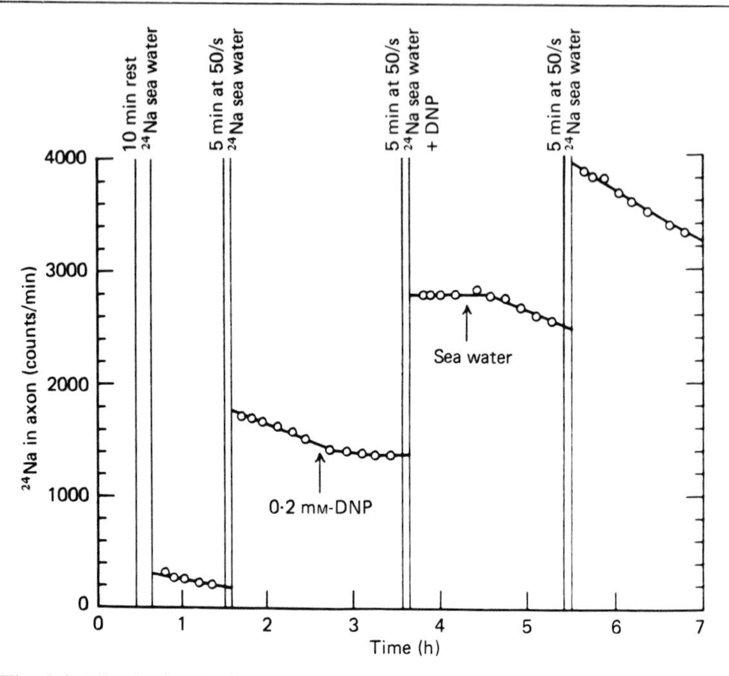

Fig. 3.9. The lack of effect of dinitrophenol on the sodium entry during stimulation of a squid axon. The resting sodium influx for the first period of immersion in ^{24}Na sea water was 50 pmole/cm^2. Temperature 17 °C. From Hodgkin and Keynes (1955).

results of an experiment in which the sodium influx into a squid giant axon was measured by soaking it for a few minutes in a solution containing ^{24}Na and then mounting it above a Geiger counter in a stream of unlabelled artificial sea water. While sodium was being actively extruded from the axon, the counting rate fell steadily, but on the addition of DNP to the sea water bathing the axon, the counts remained constant. When the DNP was washed away, the sodium pump started up again. The rate of gain of radioactivity during the periods of exposure to ^{24}Na was increased by a factor of about 10 by stimulation at 50 shocks/s, but the extra entry of ^{24}Na was the same whether or not the sodium pump had been blocked. Washing out experiments showed that the accelerated outward movement of ^{24}Na during the impulse (see p. 47) was affected equally little by DNP.

It follows from this evidence and many other considerations

Table 3.3. *Comparison of the properties of the sodium and potassium channels with those of the sodium pump*

	Sodium and potassium channels	Sodium pump
Direction of ion movements	Down the electrochemical gradient	Against the electrochemical gradient
Source of energy	Pre-existing concentration gradient	ATP
Voltage dependence	Regenerative link between potential and sodium conductance	Independent of potential
Blocking agents	Tetrodotoxin blocks Na channels at 10^{-8} M	Tetrodotoxin has no effect
	Tetramethylammonium blocks K channels at 10^{-3} M	Tetramethylammonium has no effect
	Ouabain has no effect	Oubain blocks at 10^{-7} M
External calcium	Increase in [Ca] raises threshold for excitation; decrease in [Ca] lowers threshold	No effect
Selectivity	Li^+ is not distinguished from Na^+	Li^+ is pumped much more slowly than Na^+
Effect of temperature	Rate of opening and closing of channels has large temperature coefficient, but maximum conductances have a small one	Velocity of pumping has a large temperature coefficient
Density of distribution in the membrane	Squid axon has 290 TTX-binding sites per μm^2 Rabbit vagus has 100 TTX-binding sites per μm^2	Squid axon has 4000 ouabain-binding sites per μm^2 Rabbit vagus has 750 ouabain-binding sites per μm^2
Maximum rate of movement of Na^+	100 000 pmol/cm^2 s during rising phase of action potential	60 pmole/cm^2 s at room temperature
Metabolic inhibitors	No effect; electrical activity is normal in axon perfused with pure salt solution	1 mM-cyanide or 0.2 mM-dinitrophenol block as soon as ATP is exhausted

summarized in Table 3.3 that, as indicated diagrammatically in Fig. 3.10, the pathways for the active and passive transport of ions across the membrane function quite independently of one another. This can be demonstrated most clearly in giant axons because of their large volume to surface ratio. Perhaps the most striking example of the independence of the pump and spike mechanisms was provided by Baker, Hodgkin and Shaw when they showed that a squid axon

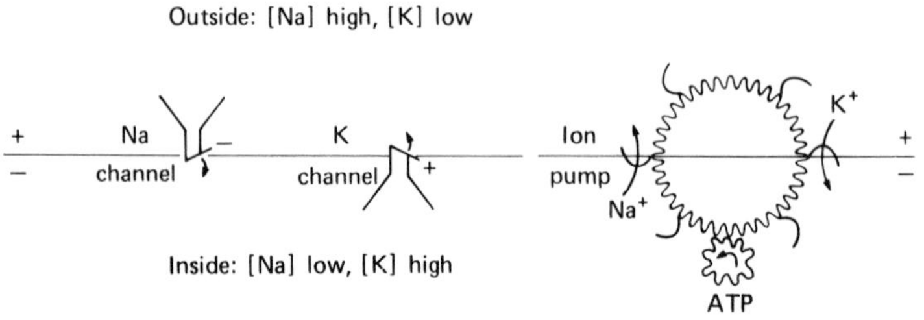

Fig. 3.10. There are two types of ionic channel traversing the nerve membrane. The sodium pump responsible for transporting ions uphill and so creating the concentration gradients is shown as a bucket system driven by ATP. The sodium and potassium channels involved in excitation are shown as funnel-shaped structures whose opening is controlled by the electric field across the membrane. In this diagram they are in the resting state with the charged gates held closed by the membrane potential. On depolarization of the membrane the gates open and permit ions to flow downhill.

whose axoplasm had been extruded and replaced by a pure solution of potassium sulphate was nevertheless capable of conducting over 400 000 impulses before becoming exhausted. In a small non-myelinated nerve fibre the downhill ionic movements during the nervous impulse are much larger in relation to the reservoir of ions built up by the sodium pump, so that blockage of active transport does, after a relatively short while, affect the conduction mechanism indirectly by reducing the size of the ionic concentration gradients.

CHAPTER FOUR

Membrane permeability changes during excitation

THE IMPEDANCE CHANGE DURING THE SPIKE

An important landmark in the development of theories about the mechanism of conduction was the demonstration by Cole and Curtis in 1939 that the passage of an impulse in the squid giant axon was accompanied by a substantial drop in the electrical impedance of its membrane. The axon was mounted in a trough between two plate electrodes connected in one arm of a Wheatstone bridge circuit (Fig. 4.1) for the measurement of resistance and capacitance in parallel. The output of the bridge was displayed on a cathode-ray oscilloscope, and R_v and C_v were adjusted to give a balance, and therefore zero output, with the axon at rest. When the axon was stimulated at one end, the bridge went briefly out of balance (Fig. 4.2) with a time course very similar to that of the action potential. The change was shown to be due entirely to a reduction in the resistance of the membrane from a resting value of about 1000 ohm cm^2 to an active one in the neighbourhood of 25 ohm cm^2. The membrane capacitance of about 1 μF/cm^2 did not alter measurably.

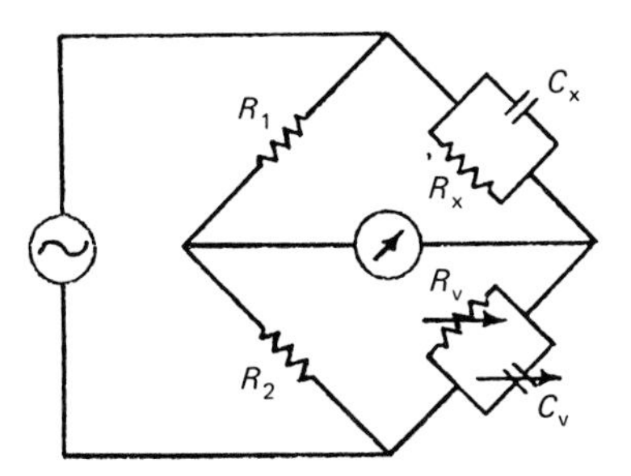

Fig. 4.1. Wheatstone bridge circuit used for the measurement of resistance and capacitance in parallel.

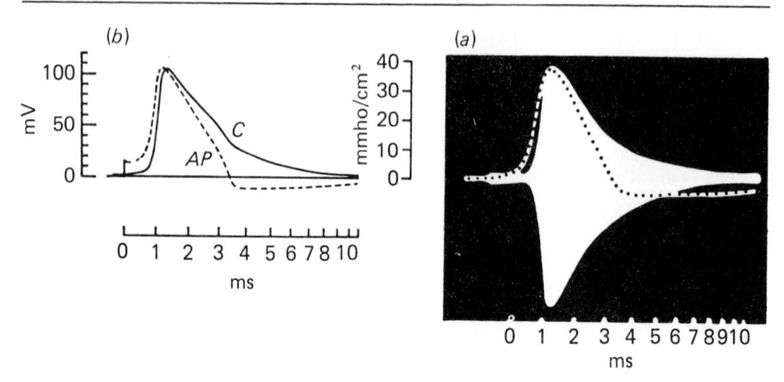

Fig. 4.2. The time course of the impedance change during the conducted action potential in a squid giant axon recorded by Cole and Curtis (1939). *a*, double exposure of the unbalance of the impedance bridge and of the monophasic action potential at one of the impedance electrodes; the time marks at the bottom are 1 ms apart. *b*, superimposed plots of the membrane conductance increase (*C*) and of the action potential (*AP*) after correction for amplifier response.

THE SODIUM HYPOTHESIS

Cole and Curtis's results were not wholly unexpected, because it had long been supposed that there was some kind of collapse in the selectivity of the membrane towards K^+ ions during the impulse. However, a year or two later both they and Hodgkin and Huxley succeeded in recording internal potentials for the first time, and it became apparent that, as has been seen in Fig. 2.4, the membrane potential did not just fall towards zero at the peak of the spike, but instead was reversed by quite a few mV. This unexpected overshoot could not possibly be accounted for by any hypothesis involving a reduction in the ionic selectivity of the nerve membrane, but required a radically different type of explanation.

None was forthcoming until in 1949 Hodgkin and Katz put forward the *sodium hypothesis* of nervous conduction. Noting that because the external sodium concentration $[Na]_o$ is greater than the internal concentration $[Na]_i$, the Nernst equilibrium potential for sodium (E_{Na}) is reversed in polarity compared with E_K, they suggested that excitation involves a rapid and highly specific increase in the permeability of the membrane to Na^+ ions, which shifts the membrane potential from its resting level near E_K to a new value that approaches E_{Na}. The first piece of evidence in support of this theory was the fact that nerves are indeed rendered inexcitable by sodium-free solutions. As Overton showed long ago for frog muscle, only Li^+ ions can fully replace Na^+, though it is now known that there are several small organic cations like hydroxylamine which can act as partial sodium substitutes; and certain excitable tissues have a calcium-dependent spike mechanism. As may be seen in Fig. 4.3,

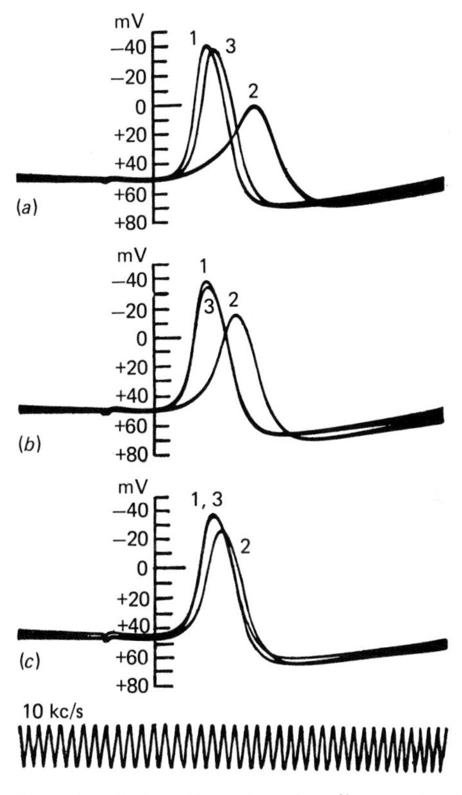

Fig. 4.3. The effect of reducing the external sodium concentration on the action potential in a squid giant axon. In each set of records, record 1 shows the response with the axon in sea water, record 2 in the experimental solution, and record 3 in sea water again. The solutions were prepared by mixing sea water and an isotonic dextrose solution, the proportions of sea water being *a*, 33%; *b*, 50%; *c*, 71%. From Hodgkin and Katz (1949).

replacement of part of the external sodium by glucose reduced both the rate of rise of the action potential and its height. The rate of rise was directly proportional to $[Na]_o$, while in accordance with eqns (3.2) and (3.3) the slope of the line relating spike height to $\log_{10} [Na]_o$ was close to 58 mV until the point was reached where conduction failed. Subsequent experiments have shown that a similar relation holds good when $[Na]_i$ is varied.

In order to change the potential across a membrane whose capacitance is 1 $\mu F/cm^2$, from -60 mV at rest to $+50$ mV at the peak of the spike, the total quantity of charge transferred must be 110 nanocoulombs/cm^2, which would be carried by 1.1 picomoles of a monovalent ion crossing 1 cm^2 of membrane. A crucial test of the validity of the sodium hypothesis was therefore to measure the net entry of sodium into the fibre and the net loss of potassium from it during the passage of an impulse. Using the technique of radioactivation analysis, Keynes and Lewis (1951) found that in stimulated *Sepia* axons there was a net gain of 3.8 pmole Na/cm^2 impulse and a net loss of 3.6 pmole K/cm^2 impulse, while in squid axons the corresponding figures were 3.5 pmole Na and 3.0 pmole K. The measured ionic movements were thus more than large enough to comply with the theory. It was not surprising that they were actually somewhat greater than the theoretical minimum, because it was reasonable to expect that there might be some exchange of potassium for sodium over the top of the spike in addition to the net uptake of sodium during its upstroke and the net loss of potassium during its falling phase. Experiments with ^{24}Na like that illustrated in Fig. 4.4 showed that there was an analogous exchange of labelled sodium during the spike as well as a net entry, for the extra inward movement of radioactive sodium was estimated as 10 pmole/cm^2 impulse, and the extra outward movement as about 6 pmole/cm^2 impulse, the difference between the two figures being in good agreement with the analytical results.

The essential new property of the membrane envisaged by the sodium hypothesis was its possession of voltage-sensitive mechanisms providing appropriate control of its sodium and potassium permeabilities. The sequence of events supposed to occur during the action potential may be summarized as follows: when the membrane

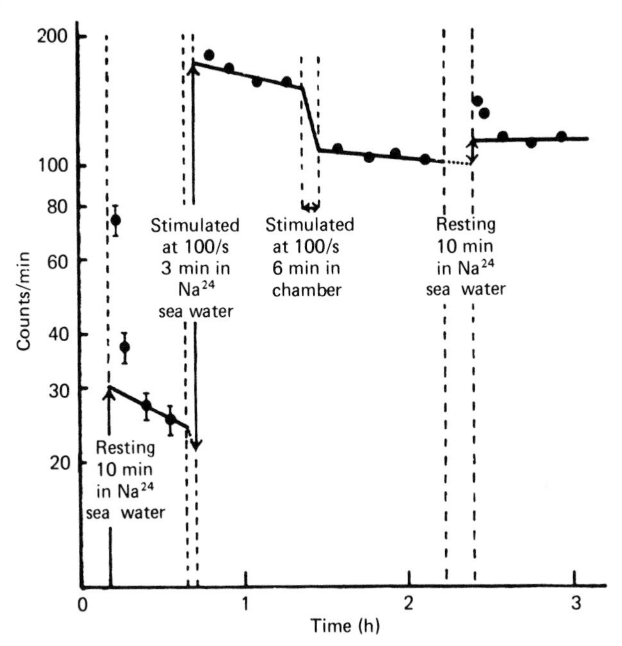

Fig. 4.4. The movements of ^{24}Na in a stimulated *Sepia* axon whose diameter was 170 μm. The axon was alternately exposed to artificial sea water containing ^{24}Na, and mounted in a stream of inactive sea water above a Geiger counter for measurement of the amount of radioactivity taken up. The loss of counts during the first ten minutes after exposure to ^{24}Na resulted from washing away extracellular sodium, and was ignored. For the entry of ^{24}Na, 1 count/min was equivalent to 42.5×10^{-12} mole Na/cm axon. Temperature 14°C. From Keynes (1951).

is depolarized by an outward flow of current, caused either by an applied cathode or by the proximity of an active region where the membrane potential is already reversed, its sodium permeability immediately rises, and there is a net inward movement of Na^+ ions, flowing down the sodium concentration gradient. If the initial depolarization opens the sodium channels far enough, Na^+ enters faster than K^+ can leave, and this causes the membrane potential to drop still further. The extra depolarization increases the sodium permeability even more, accelerating the change of membrane potential in a regenerative fashion. The linkage between sodium permeability and membrane potential forms, as shown in Fig. 4.5, a

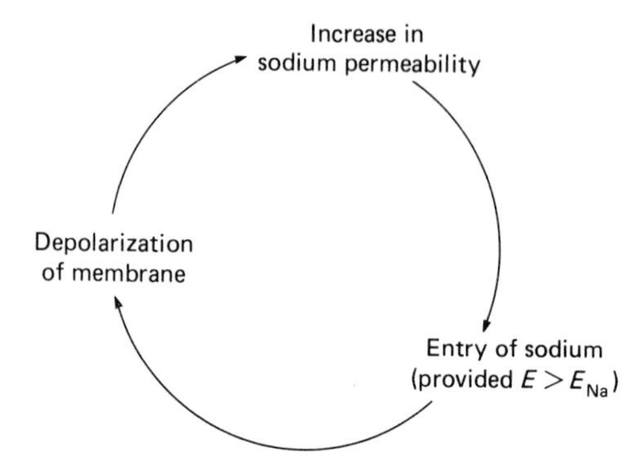

Fig. 4.5. The regenerative linkage between membrane potential and sodium permeability. From Hodgkin (1951).

positive feed-back mechanism. The entry of Na^+ does not continue indefinitely, being halted partly because the membrane potential soon reaches a level close to E_{Na}, where the net inward driving force acting on Na^+ ions becomes zero, and partly because the rise in sodium permeability decays inexorably with time from the moment when it is first triggered, this process being termed *inactivation*. After the peak of the spike has been reached, the sodium channels thus begin to close, and the sodium permeability is soon completely inactivated. At the same time, the potassium permeability of the membrane rises well above its resting value, and an outward movement of K^+ takes place, eventually restoring the membrane potential to its original level. At the end of the spike the membrane has returned to the normal resting potential, but its sodium permeability mechanism is still inactivated. Lapse of further time allows the sodium permeability to be reactivated, and hence restored to the quiescent state in which it is still very low, as is characteristic of the resting membrane, but is now ready once more to increase explosively if the system is retriggered.

According to this scheme, the most important features of the sodium channels are first that their opening is rapid and very steeply dependent on membrane potential, so that a relatively small degree of depolarization suffices to bring about a large rise in sodium permeability (P_{Na}), and second that having opened quickly they are subject to a somewhat slower process of inactivation which closes them again even though the potential has not returned to its starting level, and may still be reversed. At least in the squid giant axon, the potassium channels are controlled just as strongly by the membrane potential, but their opening is delayed and they are not inactivated, the return of P_K to normal being wholly dependent on the re-polarization of the membrane during the falling phase of the spike. The separation in time of the permeability changes, P_{Na} rising quickly and then being cut off by inactivation, while P_K only rises with an appreciable lag, helps to ensure that there is not too great an energetically wasteful interchange of Na^+ and K^+ at the peak of the spike unaccompanied by a useful alteration in the membrane potential. It may be noted that it is not essential to the conduction mechanism that P_K should increase at all. In the squid axon, the delayed rise of P_K enables the potential to return to normal faster than it otherwise would do, and so shortens the spike and speeds up conduction. But the inactivation of P_{Na} in conjuction with an unenhanced outflow of K^+ ions would bring back the potential, albeit more slowly, and some types of nerve fibre are able to dispense with the rise of P_K.

VOLTAGE-CLAMP EXPERIMENTS

The increase in the sodium permeability of the membrane during the spike that is predicted by the sodium hypothesis can be measured with radioactive tracers by the method illustrated in Fig. 4.4. But although this approach has the advantage of specificity, in that it provides unambiguous information about sodium movements and not those of any other ion, the time resolution of tracer experiments is rather poor, and the results refer only to the cumulative effect of a large number of impulses. In order to make a detailed study of the changes in membrane permeability in the course of a single action potential, it is necessary to resort to measurements of the electric current carried by the ions when they move across the membrane, which enable much greater sensitivity and much better time re-solution to be achieved. However, the amount that can be learnt simply by recording the current that flows during the conducted action potential is very limited, because the permeability changes follow a fixed sequence determined by the nerve and not by the experimenter. To get round the difficulty, Hodgkin and Huxley exploited the approach originally due to Cole and Marmont in order to measure the ionic conductance of a nerve membrane whose potential was first 'clamped' at a chosen level and then subjected to a predetermined series of abrupt changes. This enabled them to explore in considerable detail the laws governing the voltage-sensitive behaviour of the sodium and potassium channels, and present-day knowledge of the permeability mechanisms that underlie not only excitation and conduction in nerve and muscle, but also synaptic transmission, is derived very largely from *voltage-clamp* studies.

A typical experimental set-up for voltage-clamping a squid giant axon is shown in Fig. 4.6. It requires the introduction of two internal electrodes, one for monitoring the potential at the centre of the stretch of axon to be clamped, and the other for passing current uniformly across the membrane over a somewhat greater length. In Hodgkin and Huxley's original apparatus, these electrodes were of the type seen in Fig. 2.2, and were constructed by winding two spirals of AgCl-coated silver wire on a fine glass rod. Nowadays, the potential is generally recorded by a 50 μm micropipette filled with isotonic KCl (see Fig. 2.1), to which is glued a platinum wire 75 μm in diameter whose terminal portion is left bare and platinized so that it will pass current without undue polarization. The external electrode consists of a platinized platinum sheet in three sections: the current flowing to the central section is amplified and recorded, while the two outer sections help to ensure the uniformity of clamping over the fully controlled region. After appropriate amplification, the internal

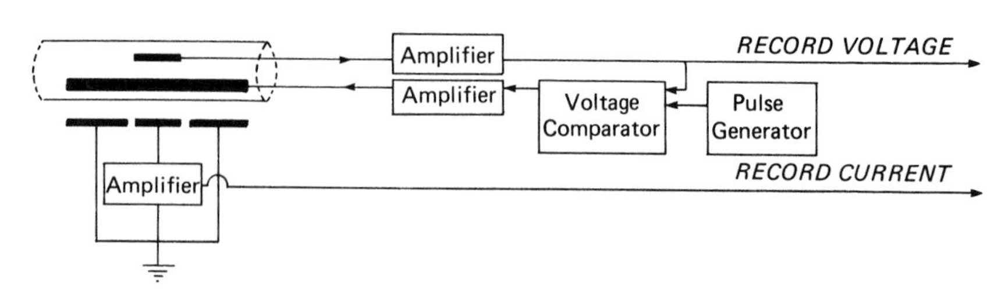

Fig. 4.6. Schematic diagram of the arrangement for measuring membrane current under a voltage-clamp in a squid giant axon.

potential is fed to a voltage comparator circuit along with the square wave signal to which it is to be clamped. The output from the comparator is applied to the internal current wire so as to increase or decrease the membrane current just enough to force the membrane potential to follow the square wave exactly. In electronic terms, this arrangement constitutes a negative feedback control system in which the potential across the membrane is determined by the externally generated command signal, and the resulting membrane current is measured. In order to voltage-clamp smaller non-myelinated nerve fibres, single nerve cells, muscle fibres, or the isolated node of Ranvier in a myelinated nerve fibre, various other electrode arrangements are called for, but the basic principle of the circuit remains the same.

The equivalent electrical circuit of the nerve membrane may be regarded, as shown in Fig. 4.7, as a capacitance C_m connected in parallel with three resistive ionic pathways each incorporating a

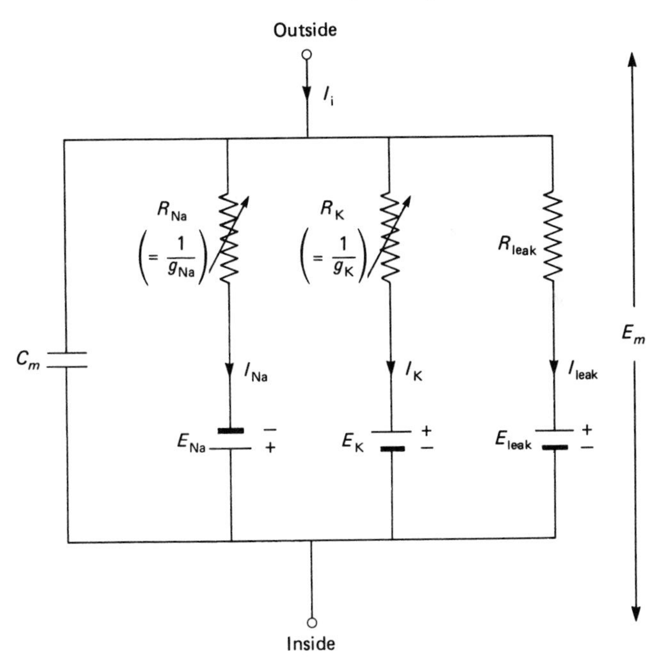

Fig. 4.7. The equivalent electrical circuit of the nerve membrane according to Hodgkin and Huxley (1952). R_{Na} and R_K vary with membrane potential and time; the other components are constant.

resistance (R_K, R_{Na} and R_{leak}) in series with a battery. For a given ionic pathway, the driving forces acting on the ions are the membrane potential E_m and the concentration gradient for that species of ion. As has been explained on p. 32, the concentration gradient may be equated with an electromotive force calculated from eqn (3.2) as the Nernst equilibrium potential, whence the appropriate values for the three battery potentials are E_K, E_{Na} and E_{leak}, and the net e.m.f. acting on each ion is the difference between E_m and its Nernst potential. It follows from Ohm's Law that the ionic currents I_K, I_{Na} and I_{leak} are given by

$$I_K = \frac{E_m - E_K}{R_K} \tag{4.1}$$

and so on. Although in accordance with electrical convention the ionic pathways are represented as resistances, it is often more convenient to think of them as the reciprocal conductances g_K, g_{Na} and g_{leak}. These represent the ease with which that particular ion can pass across the membrane, and are thus directly comparable with the permeability coefficients that appear in the constant field equation (see p. 33), though they are measured in different units. In the equivalent circuit, R_K $(=\frac{1}{g_K})$ and R_{Na} $(=\frac{1}{g_{Na}})$ are indicated as being variable, and the object of voltage-clamp experiments is to investigate their dependence on membrane potential and time. R_{leak}, to which the main contributing ion is Cl^-, is constant. In the absence of externally applied current, the electrical model predicts that the value of E_m will be determined by the relative sizes of the ionic conductances. If g_K is, as in the resting condition, much larger than g_{Na}, E_m will lie close to E_K; but when the sodium channels are opened and g_{Na} rises, E_m will move towards E_{Na}.

When the potential at which the membrane is clamped is suddenly altered, the current flowing across the membrane will consist of the capacity current required to charge or discharge C_m plus the ionic current that is to be measured. Hence the total current I will be given by

$$I = C_m \cdot \frac{dE}{dt} + I_i \tag{4.2}$$

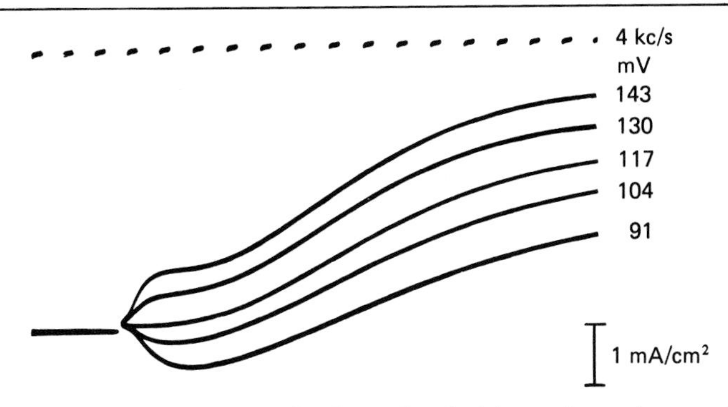

Fig. 4.8. Membrane currents for large depolarizing voltage-clamp pulses; outward current upwards. The figures on the right show the change in internal potential in mV. Temperature 3.5 °C. From Hodgkin (1958) after Hodgkin, Huxley and Katz (1952).

where I_i is the sum of the current flowing through all three ionic pathways. With a well-designed voltage-clamp system, $\frac{dE}{dt}$ should have fallen to zero and the flow of capacity current should therefore have ceased after no more than about 20 μs, so that all subsequent changes in the recorded current can be attributed to alterations in the sodium and potassium conductances operative at the new membrane potential. Fig. 4.8 shows a typical family of superimposed current records for a squid giant axon subjected to increasingly large voltage steps in the depolarizing direction. The initial capacity transients were too fast to be photographed, and what is seen is purely the ionic current. It is evident that there is an early phase of ionic current which flows inwards for small depolarizations and outwards for large ones, and a late phase which is always outwards. This is consistent with the postulates of the sodium hypothesis, and next we have to consider how the contributions of I_{Na} and I_K can be separated from one another.

The method adopted by Hodgkin and Huxley for the analysis of their voltage-clamp records was to suppress the inward sodium current by substituting choline for sodium in the external medium. This procedure yielded records of the type shown in Fig. 4.9, from which it is apparent that the removal of external sodium converts the initial hump of inward current into an outward one, but has no effect

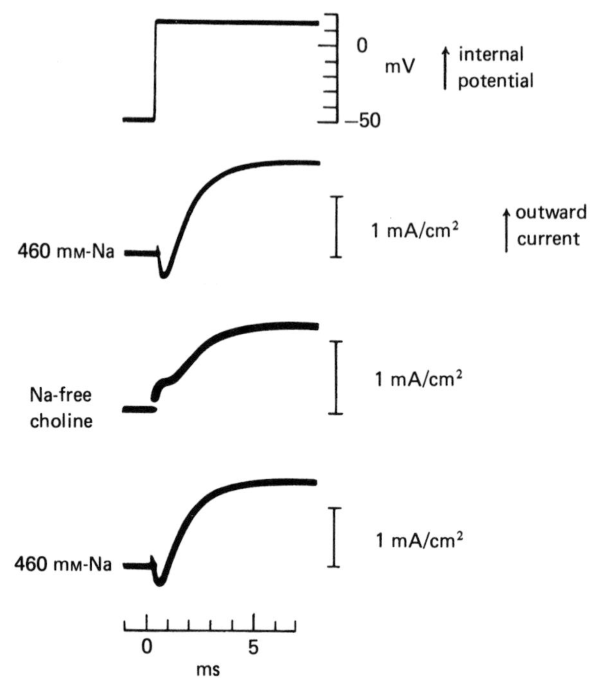

Fig. 4.9. Membrane currents associated with depolarization of 65 mV in presence and absence of external sodium. Outward current and internal potential shown upward. Temperature 11 °C. From Hodgkin (1958) after Hodgkin and Huxley (1952).

on the late current, confirming that they are carried by Na^+ and K^+ ions respectively. To eliminate the sodium current completely, it was necessary to leave some sodium in the external medium and to take E_m exactly to E_{Na}, where by definition $I_{Na}=0$. In the experiment of Fig. 4.10 this was achieved by reducing $[Na]_o$ to one tenth and depolarizing by 56 mV; trace *b* shows the resulting record of the potassium current by itself. Subtraction of trace *b* from trace *a*, which was recorded in normal sea water, then yielded trace *c* as the time course of the sodium current. The currents thus measured were finally converted into conductances by taking $g_K = I_K/(E_m - E_K)$ and $g_{Na} = I_{Na}/(E_m - E_{Na})$. A plot of g_K and g_{Na} against time (Fig. 4.11) shows that, as explained on p. 48, g_{Na} rises quickly and is then inactivated, while g_K rises with a definite lag and is not inactivated.

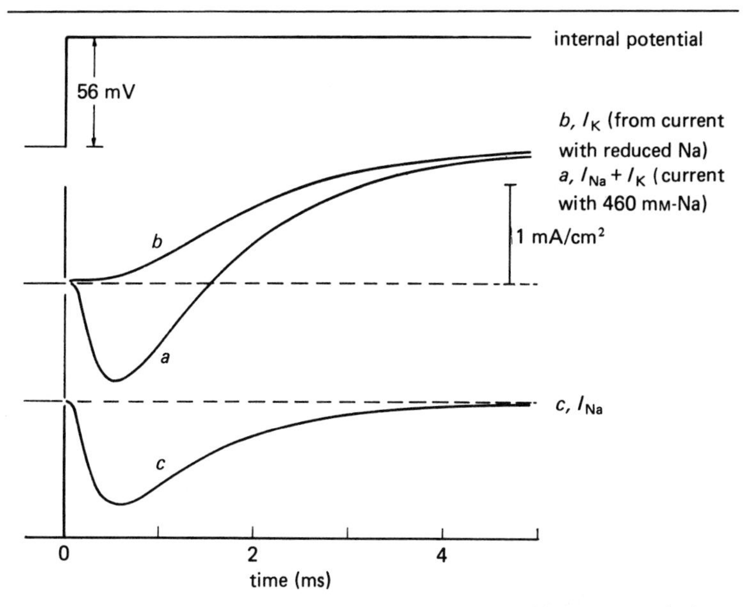

Fig. 4.10. Analysis of the ionic current changes in a squid giant axon during a voltage-clamp pulse that depolarized it by 56 mV. Trace $a(=I_{Na}+I_K)$ shows the response with the axon in sea water containing 460 mM-Na. Trace b $(=I_K)$ is the response with the axon in a solution made up of 10% sea water and 90% isotonic choline chloride solution. Trace c $(=I_{Na})$ is the difference between traces a and b. Temperature 8.5 °C. From Hodgkin (1958) after Hodgkin and Huxley (1952).

Separation of the two components of the ionic current can now be achieved more easily by recording the sodium current after completely blocking the potassium channels, and vice versa. A good method of abolishing I_K is through the introduction of caesium into the axon by perfusion or dialysis: the Cs^+ ions enter the mouths of the potassium channels from the inside, and block them very effectively. Fig. 4.12 shows a typical family of I_{Na} records for voltage-clamp pulses of different sizes applied to a squid giant axon dialysed with caesium fluoride. For the sodium channels, a blocking agent is now available which acts externally at extremely low concentrations, this being the Japanese puffer-fish-poison tetrodotoxin, usually abbreviated to TTX, whose affinity constant for the sodium sites is no more than about 3 nanomolar. Fig. 4.13 shows a family of I_K records for a squid axon dialysed with a potassium fluoride solution and

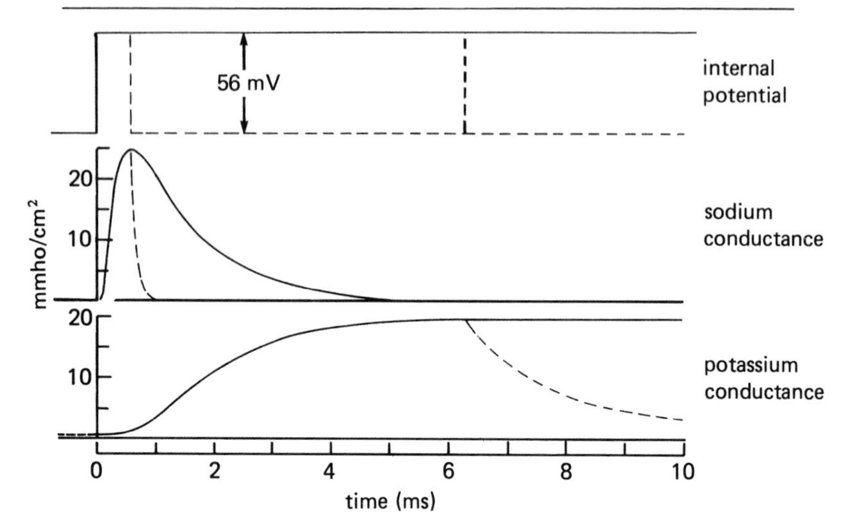

Fig. 4.11. Time courses of the ionic conductance changes during a voltage-clamp pulse calculated from the current records shown in Fig. 4.10. The dashed lines show the effect of repolarization after 0.6 or 6.3 ms. From Hodgkin (1958) after Hodgkin and Huxley (1952).

bathed in a sodium-free solution containing 1 μM-TTX. A quantitative analysis of such records gives results identical with those obtained by Hodgkin and Huxley, and not only confirms their conclusions in every respect but also provides convincing evidence for the validity of the assumption that the sodium and potassium channels are entirely separate entities between whom the only connection is a strong dependence on a potential gradient common to both of them.

Hodgkin and Huxley next proceeded to devise a set of mathematical equations which would provide an empirical description of the behaviour of the sodium and potassium conductances as a function of membrane potential and time. Thus the sodium conductance was found to obey the relationship

$$g_{Na} = \bar{g}_{Na} m^3 h \tag{4.3}$$

where $\bar{g}_{Na}$ is a constant representing the peak conductance attainable, m is a dimensionless activation parameter which varies between 0 and 1, and h is a similar inactivation parameter which varies between 1

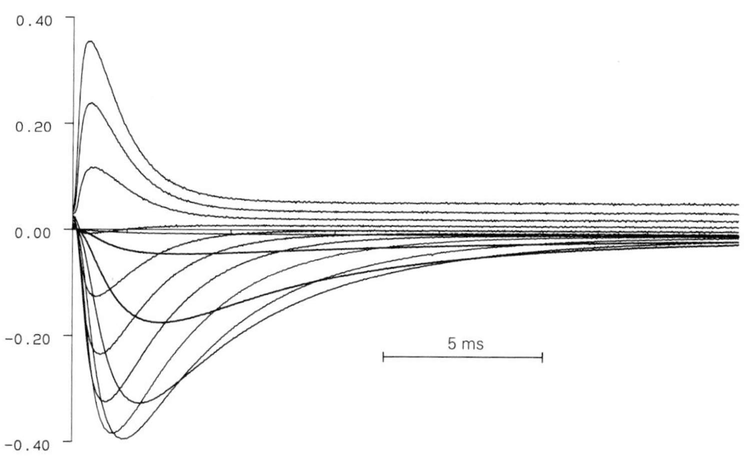

Fig. 4.12. Superimposed traces of the sodium current in a voltage-clamped squid axon whose potassium channels had been blocked by internal dialysis with 330 mM-CsF+20 mM-NaF and which was bathed in a K-free artificial sea water containing 103 mM-NaCl and 421 mM-Tris buffer. The membrane was held at −70 mV, and pulses were applied taking the potential to levels varying between −40 and +80 mV in steps of 10 mV. Current scales mA/cm². For the smaller test pulses, the current flowed inward (downward), but above about +50 mV its direction reversed. For the largest pulses, inactivation was no longer complete, and the channels ended up in the non-inactivating open state (see p. 70). Temperature 5 °C. Computer recording made by R. D. Keynes, N. G. Greeff, I. C. Forster and J. M. Bekkers.

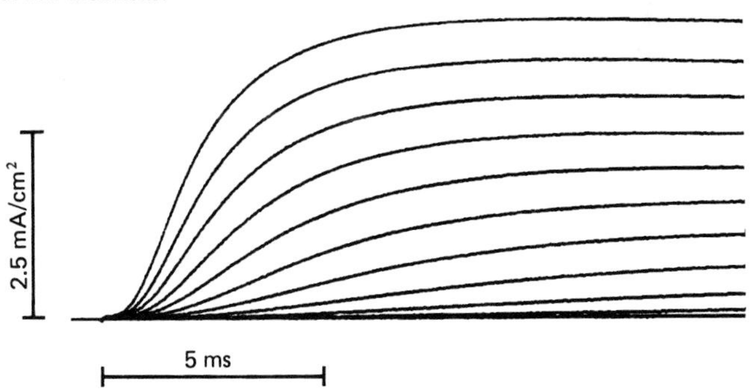

Fig. 4.13. Superimposed traces of the potassium current in a voltage-clamped squid axon whose sodium channels had been blocked by bathing it in artifical sea water containing 1 μM-TTX, and which was dialysed internally with 350 mM-KF. The membrane was held at − 70 mV, and pulses were applied taking the potential to levels varying between − 60 and + 40 mV steps. Outward current is upward. Temperature 4 °C. Computer recording made by R. D. Keynes, J. E. Kimura and N. G. Greeff.

and 0. The corresponding equation for the potassium conductance was

$$g_K = \bar{g}_K \cdot n^4 \tag{4.4}$$

where $\bar{g}_K$ is the peak potassium conductance and n is another dimensionless activation parameter. The quantities m, h and n described the variation of the conductances with potential and time, and were determined by the differential equations

$$\frac{dm}{dt} = \alpha_m(1-m) - \beta_m, \tag{4.5}$$

$$\frac{dh}{dt} = \alpha_h(1-h) - \beta_h \tag{4.6}$$

and

$$\frac{dn}{dt} = \alpha_n(1-n) - \beta_n, \tag{4.7}$$

where the αs and βs are voltage-dependent rate constants whose dimensions are time^{-1}. The precise details of the voltage-dependence of the six rate constants need not concern us further, since eqns (4.3) to (4.7) have mainly been cited in order to help the mathematically-minded reader to follow the steps that were necessary for the achievement of Hodgkin and Huxley's primary objective of testing the correctness of their description of the permeability system by calculating from their equations the shape of the propagated action potential.

The final step in Hodgkin and Huxley's arguments was thus the

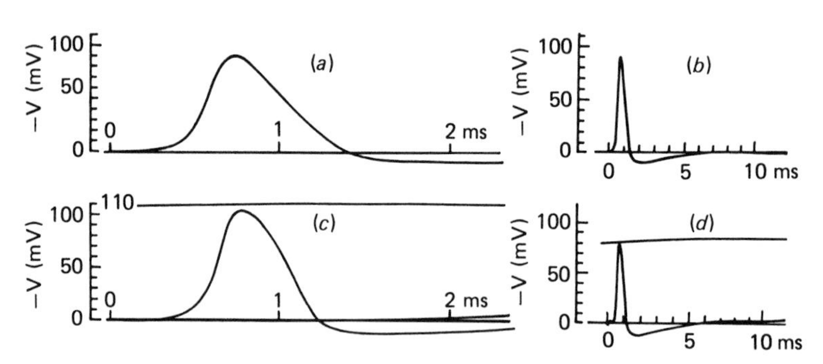

Fig. 4.14. Comparison of computed (*a,b*) and experimentally recorded (*c,d*) action potentials propagated in a squid giant axon at 18.5 °C, plotted on fast and slow time scales. The calculated conduction velocity was 18.8 m/s, and that actually observed was 21.2 m/s. From Hodgkin and Huxley (1952).

computation from data obtained under voltage-clamp conditions of the way in which the conducted action potential would be expected to behave. Fig. 4.14 shows an example of the excellent agreement between the predicted time course of the propagated action potential at 18.5 °C and what was observed experimentally at this temperature. The velocity of conduction of the impulse could also be computed, and again the theoretical and observed values were satisfactorily close to one another. Lastly, the net exchange of sodium and potassium could be predicted from the calculated extents and degree of overlap of the changes in g_{Na} and g_K during the spike that are illustrated in Figs. 4.15 and 4.16. The total entry of sodium and the exit of potassium in a single impulse each worked out to be about 4.3 pmole/cm^2 membrane, which fits very well with the results of the analytical and tracer experiments discussed on p. 46. No more could

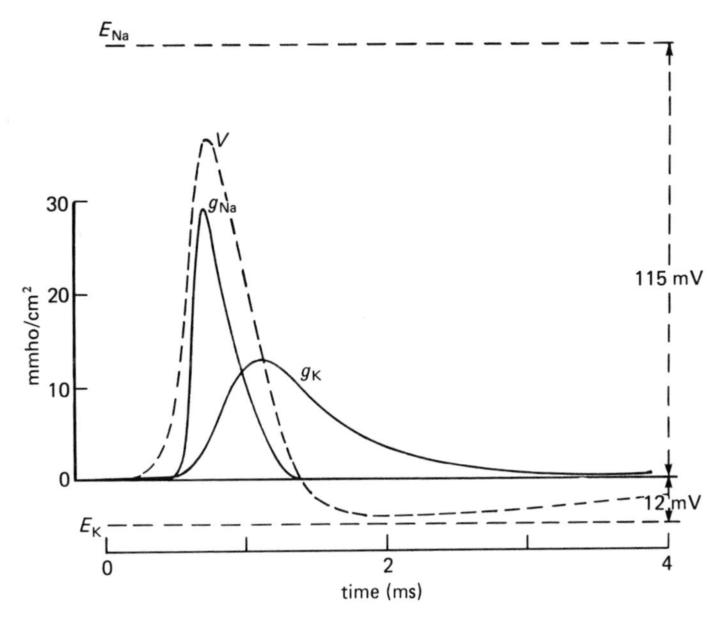

Fig. 4.15. The time courses of the propagated action potential and underlying ionic conductance changes computed by Hodgkin and Huxley from their voltage-clamp data. The constants used were appropriate to a temperature of 18.5 °C. The calculated net entry of Na^+ was 4.33 pmole/cm^2, and the net exit of K^+ was 4.26 pmole/cm^2. Conduction velocity = 18.8 m/s. From Hodgkin and Huxley (1952).

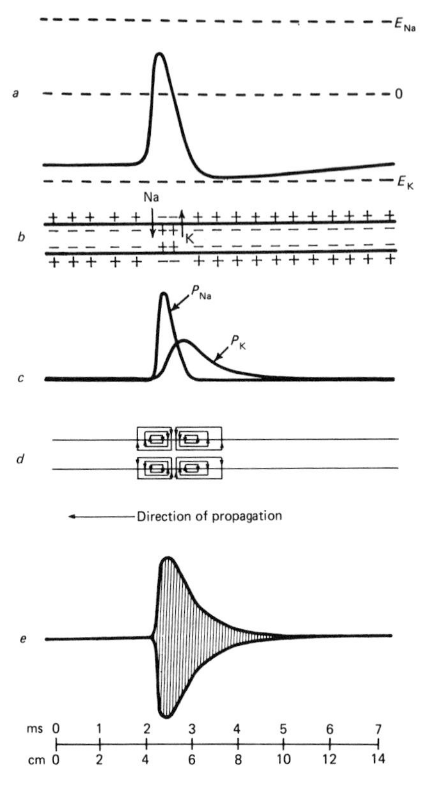

Fig. 4.16. Time relations of the events during the conducted impulse. *a*, membrane potential. *b*, ionic movements. *c*, membrane permeability. *d*, local circuit current flow. *e*, membrane impedance.

have been asked of the sodium hypothesis than that it should have yielded from purely electrical measurements figures which checked so nicely with those based on a chemical approach.

PATCH-CLAMP STUDIES

Following the introduction by Neher and Sakmann (1976) of a voltage-clamp method for observing the currents flowing through single acetylcholine-receptor channels in denervated frog muscle fibres (see p. 98), the technique of single-channel recording from a very small patch of membrane has been greatly extended (Fig. 4.17),

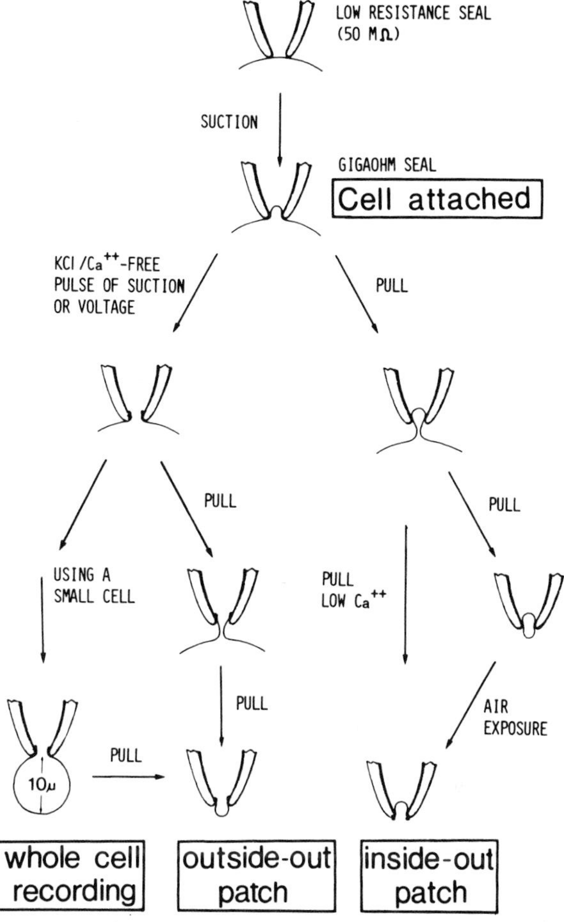

Fig. 4.17. A schematic representation of the procedures for forming a gigaohm seal between the tip of a micropipette and a patch of cell membrane, and of achieving the recording configurations known as 'cell-attached', 'whole-cell', 'outside-out patch' and 'inside-out patch'. From Hamill *et al.* (1981).

making it possible to study the properties of ion channels in every kind of living cell. The results of such studies are outside the scope of this volume, but it is clear that ion channels similar to those found in nerve and muscle have a variety of roles beyond the conduction of impulses in excitable tissues.

CHAPTER FIVE

The molecular structure and functioning of voltage-gated ionic channels

Both voltage-gated and ligand-gated ion channels are large protein molecules, as is the sodium pump Na,K-ATPase. In recent years the primary structure of a number of them has been determined, and by combining this information with the biophysical evidence major advances have been made in our understanding of how they work at the molecular and sub-molecular levels.

cDNA SEQUENCING STUDIES

A protein consists of a long chain built up of twenty different amino acids (Table 5.1), folded on itself in a rather complicated way. Its properties depend critically on the arrangement of the folds, which is determined by the exact order in which its constituent amino acids are strung together. This in turn is specified by the sequence of the nucleotide bases that make up the DNA molecules which constitute the genetic material of the cell. There are only four different bases, and each of the twenty amino acids corresponds according to a universally obeyed triplet code to a specific group of three of them. The information embodied in the base sequence of a DNA molecule is transcribed on to an intermediary messenger RNA, and is then translated during the synthesis of the protein to yield the correct sequence of amino acids. Rapid sequencing methods for nucleotides have now been perfected by Sanger and his colleagues, and modern recombinant DNA technology makes possible the cloning of DNA so that the quantity required for the determination can be prepared from a single gene. Hence the amino acid sequences of proteins are nowadays most easily determined indirectly from the base sequences of the cDNA in which they are encoded.

Table 5.1. *The amino acids found in proteins. Amino acids have the general formula R—CH(NH$_2$)COOH, where R is the side chain or residue. Proline is actually an amino acid, while cystine is two cysteines linked by a disulphide bridge. The standard abbreviations are given in three- and one-letter codes. The hydropathy index is taken from Kyte and Doolittle (1982).*

Type	Amino acid	Side chain	Abbreviations		Hydropathy index
Non-polar	Isoleucine	$-CH(CH_3)CH_2.CH_3$	Ile	I	4.5
	Valine	$-CH(CH_3)_2$	Val	V	4.2
	Leucine	$-CH_2.CH(CH_3)_2$	Leu	L	3.8
	Phenylalanine	$-CH_2.C_6H_5$	Phe	F	2.5
	Methionine	$-CH_2.CH_2.SCH_3$	Met	M	1.9
	Alanine	$-CH_3$	Ala	A	1.8
	Tryptophan	$-CH_2-$ (indole ring, N H)	Trp	W	−0.9
	Proline	$-CH_2.CH_2.CH_2-$	Pro	P	−1.6
Uncharged Polar	Cysteine/cystine	$-CH_2SH$	Cys	C	2.5
	Glycine	$-H$	Gly	G	−0.4
	Threonine	$-CH(OH)CH_3$	Thr	T	−0.7
	Serine	$-CH_2OH$	Ser	S	−0.8
	Tyrosine	$-CH_2C_6H_4OH$	Tyr	Y	−1.3
	Glutamine	$-CH_2.CH_2.CO.NH_2$	Gln	Q	−3.5
	Asparagine	$-CH_2.CO.NH_2$	Asn	N	−3.5
Acidic	Aspartic acid	$-CH_2.COO^-$	Asp	D	−3.5
	Glutamic acid	$-CH_2CH_2COO^-$	Glu	E	−3.5
Basic	Histidine	$-CH_2-$ (imidazole ring, HN NH$^+$)	His	H	−3.2
	Lysine	$-(CH_2)_4.NH_3^+$	Lys	K	−3.9
	Arginine	$-(CH_2)_3NH.C(NH_2)=NH_2^+$	Arg	R	−4.5

THE PRIMARY STRUCTURE OF THE SODIUM CHANNEL

The substantial voltages generated by the electric organ of the electric eel depend on the additive discharge of a large number of cells that are derived embryonically from muscle (see Fig. 2.4*g*). Their electrical excitability involves an increase of sodium permeability in the usual way, and this type of electric organ therefore provided ideal material first for isolating and purifying the sodium channel protein, and then for enabling its amino acid sequence to be determined. The initial biochemistry was greatly facilitated by the fact that the protein could be labelled with high specificity by the Japanese puffer-fish-poison tetrodotoxin (TTX). The sodium channel was shown to be a single large peptide with a molecular weight of about 260 kDaltons, which is glycosylated at several points on incorporation in the membrane.

A team of scientists led by Numa and Noda at Kyoto University has successfully cloned and sequenced the cDNA of the *Electrophorus* sodium channel. The protein is made up of 1820 amino acid residues, and as shown in Figs. 5.1 and 5.2, contains four homologous domains spanning the membrane which have very

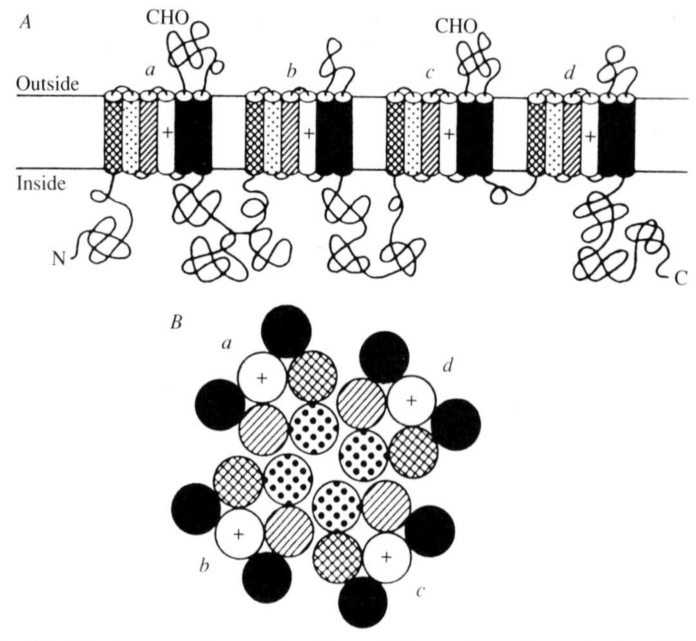

Fig. 5.1. The transmembrane topology of the sodium channel, viewed (*A*) in the plane of the membrane, and (*B*) perpendicular to it. In *A*, the homology domains *a–d* that span the membrane are displayed linearly. Segments S1–S6 in each repeat are indicated by cylinders as follows: S1, cross-hatched; S2, stippled; S3, hatched; S4, indicated by a plus sign; S5 and S6, solid. Putative sites of *N*-glycosylation (CHO) are indicated. The ionic channel appears in *B* as the central pore surrounded by the four S2 segments. After Noda *et al.* (1986).

similar amino acid sequences. The sodium channel genes in rat brain and *Drosophila* have subsequently been found to contain closely comparable homologous regions, which along with the short cytoplasmic link joining domains *c* and *d* appear to be the parts of the molecule most highly conserved through evolution.

Of the twenty possible amino acids that make up a protein, it may be seen in Table 5.1 that the residues of eight are non-polar, seven are polar but uncharged, two are acidic and carry a negative charge, and three are basic and positively charged. The non-polar residues are hydrophobic, and therefore tend to be located in the lipid core of the membrane or in the middle of the molecule. The polar or charged residues are hydrophilic, and are more likely to be found in the aqueous environment of the cytoplasm or at the surface of the membrane. From a study of the so-called hydropathicity index of the different stretches of the amino acid chain it has been deduced that each of the homologous domains contains six segments represented as cylinders in Fig. 5.1*A*, that are largely hydrophobic and form α-helices crossing the membrane from one side to the other. Numa and his colleagues suggest that the four homologous domains are arranged as in Fig. 5.1*B* so as to surround a central aqueous pore whose wall may be lined with the negative charges of the S2 segments, making it permeable to cations.

An essential requirement of the structure is that it should incorporate voltage-sensing elements that will respond to alterations in the electric field across the membrane. Although the reversal in membrane potential during the conducted impulse is only around 100 mV, this corresponds to a gradient of more than 300000 V/cm across the part of the membrane that has a high dielectric constant (see p. 28). It is generally agreed that the best candidates to act as the voltage-sensors are the S4 segments (Fig. 5.2), in which every third residue is a positively charged arginine or lysine alternating with non-polar or uncharged polar residues. According to the screw helical hypothesis (Catterall, 1986; Guy, 1988), these positive charges are paired with negatively charged residues in neighbouring segments in such a way that a twist of 60° enables the S4 α-helix to form a fresh set of pairs, leaving an unpaired negative charge at one end, and an unpaired positive charge at the other, and transferring

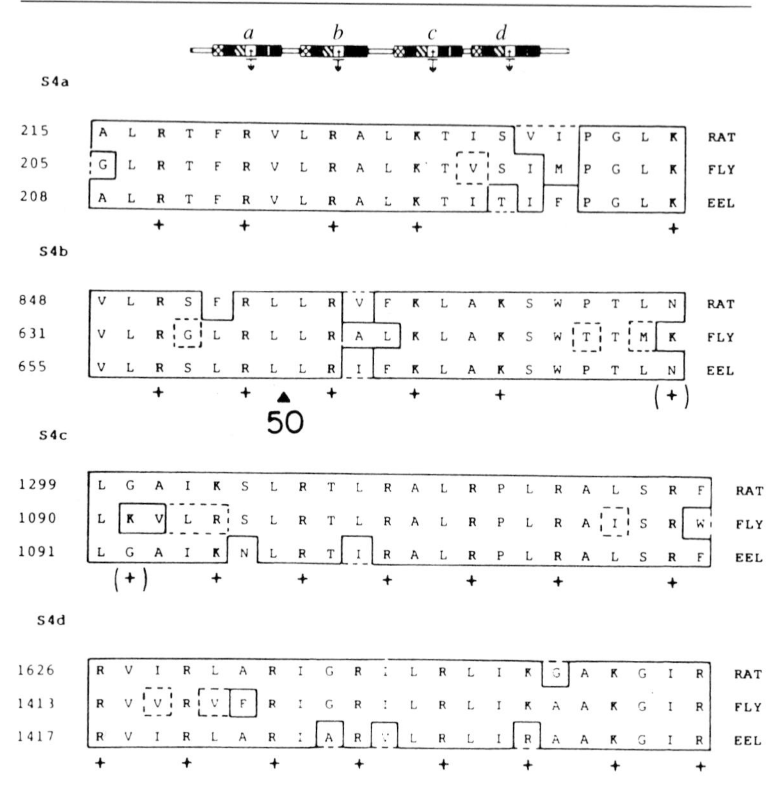

Fig. 5.2. The amino acid sequences of the S4 segment of homology domains *a*, *b*, *c* and *d* in the sodium channels from rat brain, *Drosophila* muscle, and electric eel. Solid lines enclose identical residues; dotted lines enclose conservative substitutions. Positively charged arginine (R) or lysine (K) residues are labelled with a +. The other residues, identified according to the one-letter code in Table 5.1, are uncharged. A bracket indicates a charge present only in the *Drosphila* sequence. The caret in S4b indicates the position and size of an intron. From Salkoff *et al.* (1987).

about 1.2 units of charge inwards or outwards across the membrane. The result is an allosteric conformational change in the structure of the central pore, and the generation of sodium gating current (see below).

Expression in the oocyte

In order to test the conclusions of cDNA sequencing studies, it is important to make sure that the proteins whose structures have been determined are indeed capable of functioning correctly when inserted into the membrane. The oocytes of the African clawed toad *Xenopus* have provided a most valuable approach for tackling this problem.

The oocytes are large cells which are about to develop into mature eggs. They possess the normal translation machinery, and will respond to the injection of messenger RNA by making the protein for which it codes and incorporating it in the membrane. After synthesizing the corresponding messenger RNA, the majority of the voltage-gated and ligand-gated channel proteins that have so far been isolated have been successfully expressed in *Xenopus* oocytes, and shown by recording both macroscopic and single-channel ion currents to behave in an essentially normal fashion. An important extension of this technique is to alter the sequence of amino acid residues by the procedure known as site-directed mutagenesis, so as to be able to explore in detail the effect of artificial modifications of the protein structure.

THE SODIUM GATING CURRENT

A limitation of cDNA sequencing studies is that although they provide a wealth of direct information about the primary structure of membrane proteins, we have to depend on indirect and sometimes speculative arguments to decide how the molecule is folded, and to elucidate the nature of the conformational changes that bring about the opening and closing of the channels. Observations of the macroscopic ion currents are also restricted in their scope, because they throw light only on the kinetics of the open state, and reveal relatively little about the several closed states through which the system must certainly pass during activation and inactivation.

It was pointed out by Hodgkin and Huxley (1952) that the voltage-dependence of the sodium conductance implies that the gating mechanism itself is charged, and further that whenever a change in membrane potential operates the gate, there must be a movement down the electric field of the charged side groups that it carries, giving rise to a displacement current which necessarily precedes the flow of ionic current. The *asymmetry* or *gating current*, as it has come to be called, remained undetected for some years because the corresponding transfer of charge within the membrane is so small compared with the transfer of ions across it. However, in 1972/3 Armstrong and Bezanilla, followed shortly afterwards by Keynes and Rojas, succeeded in recording an asymmetrical component of the displacement current in the squid giant axon, thus opening up a fresh approach to the gating problem that has the unique advantage of enabling direct observations to be made on the kinetics of some at least of the closed states, though some years elapsed before it was successfully exploited.

In order to record the gating current, it is first necessary to abolish the ionic current that would otherwise swamp it. The sodium channels are accordingly blocked by bathing the axon in a sodium-free solution containing a high concentration of TTX, which fortunately turns out to seal up their mouths completely without interfering at all with the operation of the gate. The potassium channels are blocked by perfusion or internal dialysis with a caesium or tetramethylammonium fluoride solution. It is also necessary to get rid of the large capacity transient that arises when charging or discharging the passive membrane capacity. This is symmetrical with respect to potential, and using a computerized recording system it can be eliminated electronically.

Although gating currents have now been recorded at the node of Ranvier and in various other types of nerve and muscle, the preparation that enables the best possible time resolution to be achieved is the squid giant axon. The most recent recordings by Keynes, Greeff and Forster (1990) have yielded families of traces like that illustrated in Fig. 5.3*a*. It has become evident that the current is made up of several kinetically distinct components, which include an early one with a very fast relaxation that produces the brief initial peak, a much larger component that also rises very quickly and relaxes with a time constant of 50–150 μs at 10 °C, and two components of intermediate size and relaxation rate. One of these again rises quickly, while the other reaches its peak with a slight delay as may be seen in Fig. 5.3*c*, producing a shoulder in the traces of Fig. 5.3*a* when large test pulses are applied. Fig. 5.3*b* shows that the slowly rising component is abolished by inactivation of the

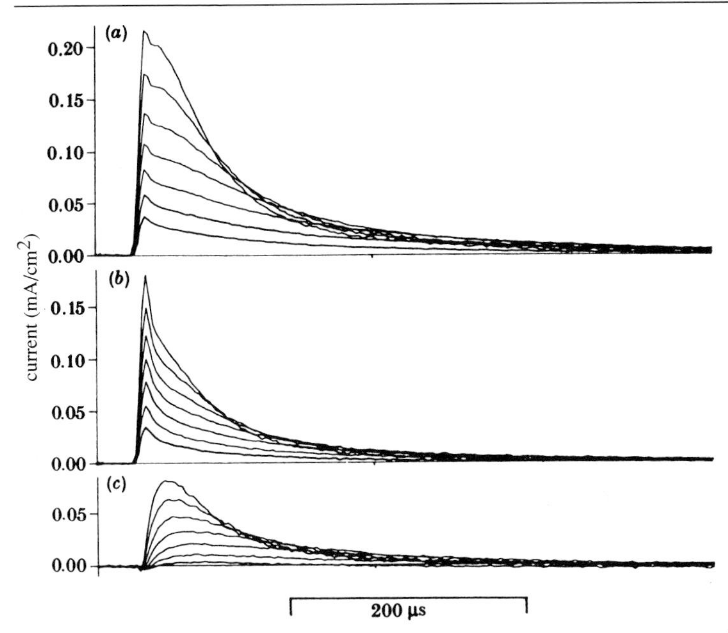

Fig. 5.3. Superimposed families of the sodium gating current in a squid giant axon dialyzed with TMAF at 10 °C. The holding potential was −80 mV, and test pulses were applied taking the potential to −40, −20, 0, 20, 40, 60 and 80 mV. Family (*a*) shows total gating current. Family (*b*) shows the result of applying the same pulses after first inactivating the sodium system with a prepulse: the component with a delayed rise has disappeared. Family (*c*) shows (*a*)−(*b*). From Keynes, Greeff and Forster (1990).

sodium system. A still slower component which relaxes in parallel with the inactivation process can also be recorded under favourable conditions.

MODELS OF THE SODIUM CHANNEL

The simplest functional model of the sodium channel suggested by Hodgkin and Huxley's equations (4.3) and (4.4) would be that there are three voltage-activated *m* particles and one inactivating *h* particle operating independently in parallel. Although at first sight these particles might be directly identified with the four S4 units, this simple interpretation has to be rejected because it fails to

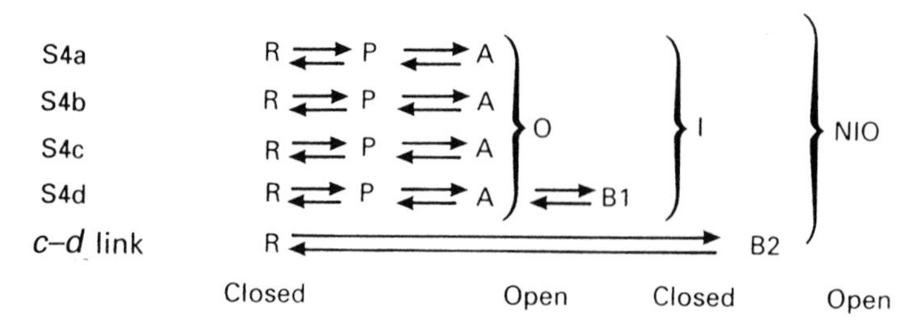

Fig. 5.4 State diagram of a series-parallel model of the sodium channel proposed by Keynes (1990, 1991). The channel opens when all four of the S4 units have been activated to bring it to the 4A (=O) state. Following a third step of S4d to state B1, a blocking group associated with the *c–d* link shuts off the channel in the closed inactivated state I. In the final state B2 (=NIO), the blocking group takes up a different position, allowing some of the channels to be reopened by large test pulses.

account for several features of the behaviour of the gating currents, and in particular for the evidence that activation and inactivation are in some way coupled sequentially. However, a detailed analysis of the properties of the kinetic components has now led to the concept of the series-parallel arrangement of the voltage-sensors illustrated in Fig. 5.4, which is in effect a refined version of the Hodgkin–Huxley model, and explains the initial delay in the opening of the channel in the same way. It proposes that each of the S4 units is activated in two stages by successive twists of the α-helix, and that when all four in a given channel reach state A, there is an allosteric change in the structure that opens up a central aqueous pore. This opening is only transitory, because a third twist of voltage-sensor S4d moves a blocking charge into position at the inner mouth that brings the channel into the closed inactivated state. A few of the channels may later reopen by passing into the non-inactivating open state (NIO) first described by Chandler and Meves (1970), which as may be seen in Fig. 4.12 becomes quite prominent when large test pulses are applied. A computer simulation shows that such a model can correctly predict both the behaviour of the gating current and the kinetics of the opening of the channels, as well as the recovery of the system after its inactivation. Although the mechanism of coupling between states

O, I and NIO that is pictured here is still speculative, there is good experimental evidence that inactivation somehow involves positive charges carried by the short cytoplasmic link connecting homology domains *c* and *d*, for inactivation is blocked by perfusion with the proteolytic enzyme pronase or by interfering with the *c–d* link in other ways.

OTHER VOLTAGE-GATED CHANNELS

Recombinant DNA techniques have also been applied to some of the other voltage-gated ion channels. Thus Numa's group in Kyoto has shown that the calcium channel protein in rabbit muscle has 1873 amino acid residues whose sequence is strikingly alike that of the sodium channel in having four homologous domains in which about 30% of the positions are occupied by identical residues, and a further 25% by ones that are conservative in nature. The S4 segment in each domain again has lysine or arginine as every third residue. This suggests strongly that the voltage-gated sodium and calcium channels share a common evolutionary origin.

The fruit fly *Drosophila* has been the subject of intensive genetic analysis since the early years of this century. The mutant *Shaker* affects the fast transient potassium currents of the flight muscles. As in certain mollusc muscles, these currents differ from those normally observed in squid axons in that they undergo a fairly rapid inactivation after depolarization, and hence are known as A currents. Recombinant DNA studies by Tempel and his colleagues have enabled the amino acid sequence of the *Shaker* locus gene product to be determined. The putative A channel protein contains 616 residues, has a molecular weight of 70 kDaltons, and includes six hydrophobic and therefore membrane-spanning segments, together with a seventh that closely resembles the S4 units of the sodium channel in having seven arginine residues and one lysine interspersed between non-polar or uncharged residues, so that there are positive charges in every third position. Such a protein would probably be too small to form a voltage-gated ion channel on its own in the manner pictured in Fig. 5.1*B*, and it seems likely that the A channel is either a homomultimer formed by an association of four copies of the *Shaker* protein, or else a heteromultimer combining the products of more than one gene similar to *Shaker*.

Although the gating current for potassium has proved more difficult to measure than the sodium gating current, the potassium channel in the squid giant axon has long been known to have one important characteristic that it shares with the sodium channel, for its opening is appreciably delayed by the application of a negative prepulse. The occurrence of such a delay was a vital clue in arriving at the model of the sodium channel shown in Fig. 5.4, and the same mechanism could be at work in both types of channel. This suggestion is strongly supported by the conclusions of Zagotta and Aldrich (1990) based on single channel recordings for the A_1-type potassium channel in *Drosophila* muscle, which have led them to a very similar model.

THE IONIC SELECTIVITY OF VOLTAGE-GATED CHANNELS

The implications of both the structural and the biophysical findings are consequently that all voltage-gated channels are fundamentally similar in their mode of operation. It has been suggested by Hille (1989) that potassium channels were the first to evolve, followed later by the appearance of channels with rather different ionic filters in the part of the molecule that allows ions to enter into the gated section.

In the case of the sodium channel, the selectivity filter is located at the outside, in the neighbourhood of the TTX binding site. From studies of the relative permeability of the sodium channel at the node of Ranvier to certain small organic cations, Hille has concluded that selection depends partly on a good fit between the dimensions of the penetrating ion and those of the mouth of the channel. As indicated in Fig. 5.5, only molecules measuring less than about 0.3 by 0.5 nm in cross-section are able to pass the filter. However, there were striking differences in permeability between some cations whose size was the same. Thus hydroxylamine ($OH—NH_3^+$) and hydrazine ($NH_2—NH_3^+$) readily entered the channel, but methylamine ($CH_3—NH_3^+$) did not. In order to explain the discrepancy, Hille proposed that the sodium channel is

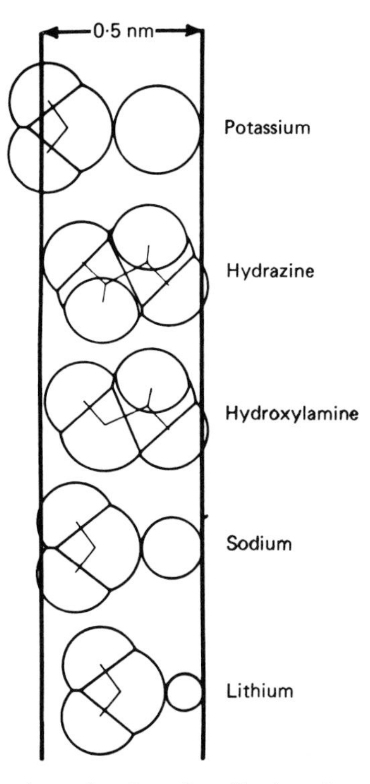

Fig. 5.5. Scale drawings showing the effective sizes of lithium, sodium and potassium ions, each with one molecule of water, and of unhydrated hydroxylamine and hydrazine ions. The vertical lines 0.5 nm apart represent the postulated space between oxygen atoms available for cations able to pass through the sodium channel. Methylamine would look just like hydrazine in this kind of picture, but is nevertheless unable to enter the channel. After Hille (1971).

lined at its narrowest point with oxygen atoms, one of which is an ionized carboxylic acid group (COO^-) bearing a negative charge. Positively charged ions containing hydroxyl (OH) or amino (NH_2) groups are able to pass through the channel by making hydrogen bonds with the oxygens, but those containing methyl (CH_3) groups are excluded from it by their inability to form hydrogen bonds. The geometry of the situation is such that Na^+ ions can divest themselves of all but one of their shell of water molecules by interacting with the strategically placed oxygen atoms, and the energy barrier that they encounter is therefore relatively low. The same is true for Li^+, but the somewhat larger K^+ ions cannot shed their hydration shell as easily, making P_K for the sodium channel only one twelfth as great as P_{Na}.

The selectivity filter of the potassium channel is located at its inner end, and operates in an analogous fashion to the extent that its structure is such that K^+ ions now encounter an appreciably lower energy barrier than more heavily hydrated Na^+ or Li^+ ions. There is evidence suggesting that there may be more than one such barrier within the channel, and that only one ion can enter it a time, giving rise to the so-called 'single file' behaviour first observed for the potassium channels of *Sepia* axons by Hodgkin and Keynes. A similar interaction between inwardly and outwardly moving ions is observed in the sodium channel, though it is much less marked.

CHAPTER SIX

Cable theory and saltatory conduction

THE SPREAD OF POTENTIAL CHANGES IN A CABLE SYSTEM

The propagation of the nervous impulse depends not only on the electrical excitability of the nerve membrane, but also on the *cable structure* of the nerve. We have already seen that the passive electrical properties of a patch of membrane can be represented as a capacitance C_m in parallel with a resistance R_m, so that the circuit diagram of a length of axon is the network shown in Fig. 6.1, where R_o is the longitudinal resistance of the external medium, and R_i is the longitudinal resistance of the axoplasm. Such a network is typical of a sheathed electric cable, albeit one with rather poor insulation, because R_m is not nearly as large compared with R_i as it would be if the conducting core were a metal. If a constant current is passed transversely across the membrane so as to set up a potential difference V_o between inside and outside at one point, then the voltage elsewhere will fall off with the distance x in the manner indicated in the lower part of Fig. 6.1. The law governing this passive *electrotonic* spread of potential is

$$V_x = V_o e^{-x/\lambda}. \tag{6.1}$$

where the *space constant* λ is given by

$$\lambda^2 = \frac{R_m}{R_o + R_i} \tag{6.2}$$

A similar argument applies in the case of a brief pulse of current, except that the value of C_m then has to be taken into account in addition to that of R_m. However, it is not necessary to enter here into the detailed mathematics of the passive spread of potential in a cable system, and it will suffice to note that theory and experiment are in good agreement.

Suppose now that an action potential of amplitude V_o has been set up over a short length of axon, and that the threshold potential

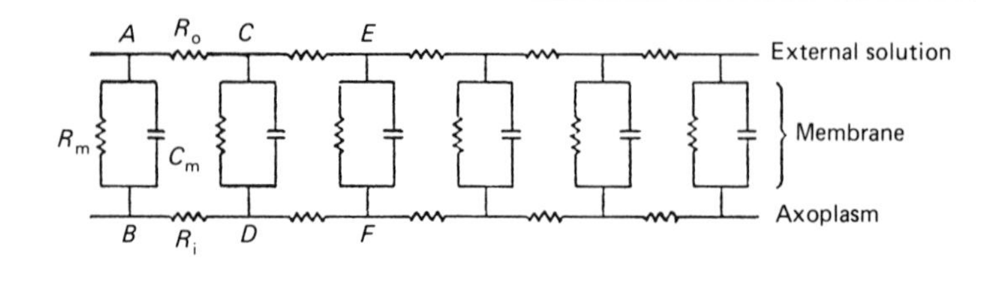

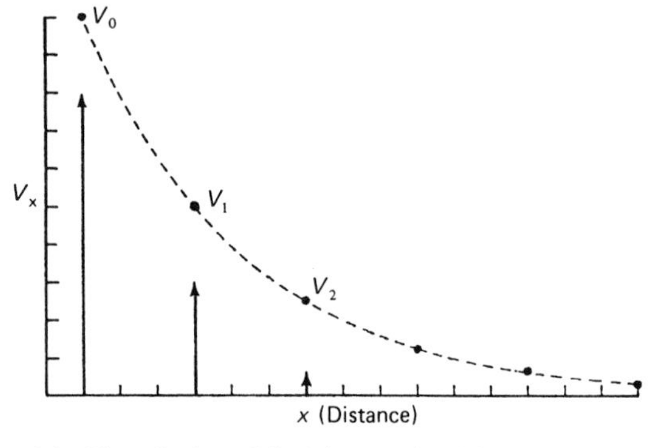

Fig. 6.1. Electrical model of the passive (electrotonic) properties of a length of axon. The graph below shows the steady-state distribution of transmembrane potential when points *A* and *B* are connected to a constant current source.

change necessary to stimulate the resting membrane is a certain fraction, say one-fifth of V_0. Because of the cable structure of the nerve, current will flow in local circuits on either side of the active region as indicated in Fig. 6.2, and the depolarization will spread passively. At a critical distance in front of the active region, which with the assumption made above would be about 1.5 times the space constant, the amount of depolarization will just exceed threshold. This part of the axon will then become active in its turn, and the active region will move forwards. Provided that the axon is uniform in diameter and in the properties of its membrane, both the amplitude and the conduction velocity of the action potential will be constant, and it will behave in an all-or-none fashion. Since the amplitude of the action potential is always much greater than the threshold for stimulation, the conduction mechanism embodies a large safety factor, and the spike can be cut down a long way by

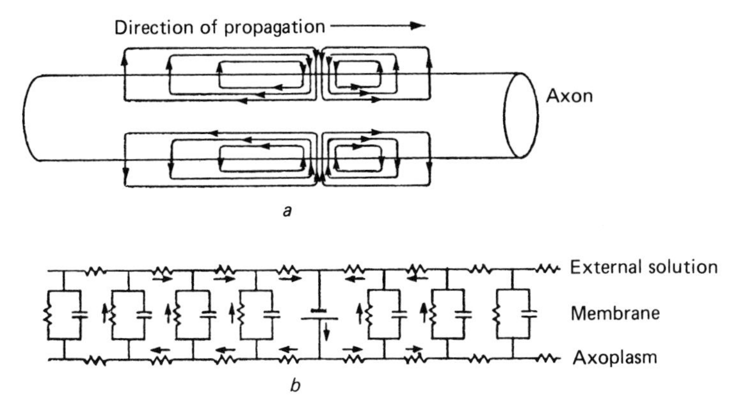

Fig. 6.2. *a*, The local circuit currents that flow during a propagated action potential; *b*, The local circuit currents set up by a battery inserted in the core-conductor model.

changes in the conditions that adversely affect the size of the membrane potential before conduction actually fails. It will be appreciated that although there is outward current flowing through the membrane both ahead of the active region and behind it, propagation can only take place from left to right in the diagram of Fig. 6.2, because the region to the rear is in a refractory state. In the living animal, action potentials normally originate at one end of a nerve, and are conducted unidirectionally away from that end. In an experimental situation where shocks are applied at the middle of an intact stretch of nerve, the membrane can of course be excited on each side of the stimulating electrode, setting up spikes travelling in both directions.

It should be clear from this description that conduction will be speeded up by an increase in the space constant for the passive spread of potential, because the resting membrane will be triggered further ahead of the advancing impulse. This is one of the reasons why large axons conduct impulses faster than small ones, for it follows from eqn (6.2) that λ is proportional to the square root of fibre diameter. Another factor that greatly affects λ is myelinization of the nerve, and we must next discuss this in more detail.

SALTATORY CONDUCTION IN MYELINATED NERVES

In 1925 Lillie suggested that the function of the myelin sheath in vertebrate nerve fibres might be to restrict the inward and outward passage of local circuit current to the nodes of Ranvier, so causing the nerve impulse to be propagated from node to node in a series of discrete jumps. He coined the term *saltatory conduction* for this kind of process, and supported the idea with some ingenious experiments on his iron wire model. (An iron wire immersed in nitric acid of the right strength acquires a surface film along which a disturbance can be propagated by local circuit action; the mechanism has several features analogous with those of nervous conduction, for which it has served as a useful model.) The hypothesis could not be tested physiologically until methods had been developed for the dissection

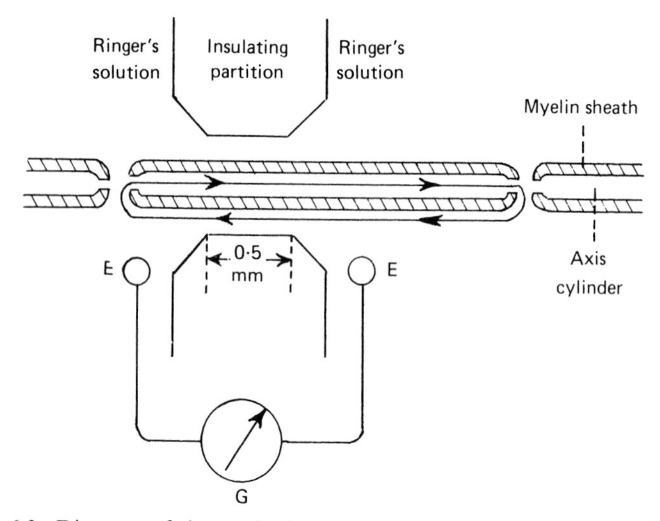

Fig. 6.3. Diagram of the method used by Huxley and Stämpfli (1949) to investigate saltatory conduction in nerve. The nerve fibre is pulled through a fine hole about 40 μm in diameter in an insulator by a micromanipulator. Current flowing along the axis cylinder out of one node and into the other as indicated by the arrows causes a voltage drop outside the myelin sheath. The resistance of the fluid in the gap between the two pools of Ringer's solution being about half a megohm, the potential difference between them can be measured by the oscilloscope *G* connected to electrodes *E* on either side. The internodal distance in a frog's myelinated nerve fibre is about 2 mm.

of isolated fibres from myelinated nerve trunks, which was first done by Kato and his school in Japan in about 1930. Ten years later Tasaki produced strong support for the saltatory theory by showing that the threshold for electrical stimulation in a single myelinated fibre was much lower at the nodes than along the internodal stretches, and that blocking by anodal polarization and by local anaesthetics was more effective at the nodes than elsewhere. In collaboration with Takeuchi, Tasaki also introduced a technique for making direct measurements of the local circuit current flowing at different positions, and this approach was subsequently perfected by Huxley and Stämpfli.

The method adopted by Huxley and Stämpfli (Fig. 6.3) was to

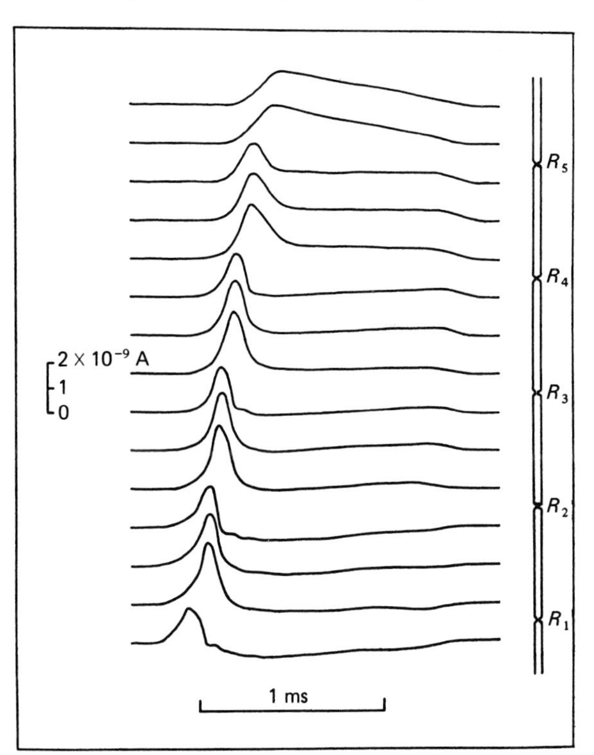

Fig. 6.4. Currents flowing longitudinally at different positions along an isolated frog nerve fibre. The diagram of the fibre on the right-hand side shows the position where each record was taken. The distance between nodes was 2 mm. From Huxley and Stämpfli (1949).

pull a myelinated fibre isolated from a frog nerve through a short glass capillary mounted in a partition between two compartments filled with Ringer's solution. The fluid-filled space around the nerve inside the capillary was sufficiently narrow to have a total resistance of about 0.5 megohm, so that the current flowing longitudinally between neighbouring nodes outside the myelin sheath gave rise to a measurable potential difference between the two sides of the partition, which could be recorded with an oscilloscope. The records of longitudinal current showed (Fig. 6.4) that at all points outside any one internode the current flow was roughly the same both in magnitude and timing. However, the peaks of current flow were displaced stepwise in time by about one tenth of a millisecond as successive nodes were traversed. In order to determine the amount of current that flowed radially into or out of the fibre, neighbouring pairs of records were subtracted from one another, since the difference between the longitudinal currents at any two points could only have arisen from current entering or leaving the axis cylinder between those points. This procedure gave the results illustrated in Fig. 6.5, from which it is seen that over the internodes there was merely a slight leakage of outward current, but that at each node there was a brief pulse of outward current followed by a much larger pulse of inward current. The current flowing transversely across the myelin sheath is exactly what would be expected for a passive leak, while the restriction of inward current to the nodes proves conclusively that the sodium system operates only where the excitable membrane is accessible to the outside.

The term 'saltatory' means literally a process that is discontinuous, but it would nevertheless be wrong to suppose that only one node is active at a time in a myelinated nerve fibre. The conduction velocity in Huxley and Stämpfli's experiments was 23 mm/ms, and the duration of the action potential was about 1.5 ms, so that the length of nerve occupied by the action potential at any moment was about 34 mm, which corresponds to a group of 17 neighbouring nodes. In the resistance network equivalent to a myelinated fibre (Fig. 6.6), the values of R_n and R_i are such that the electrotonic potential would decrement passively to 0.4 between one node and the next. Since the size of the fully developed action

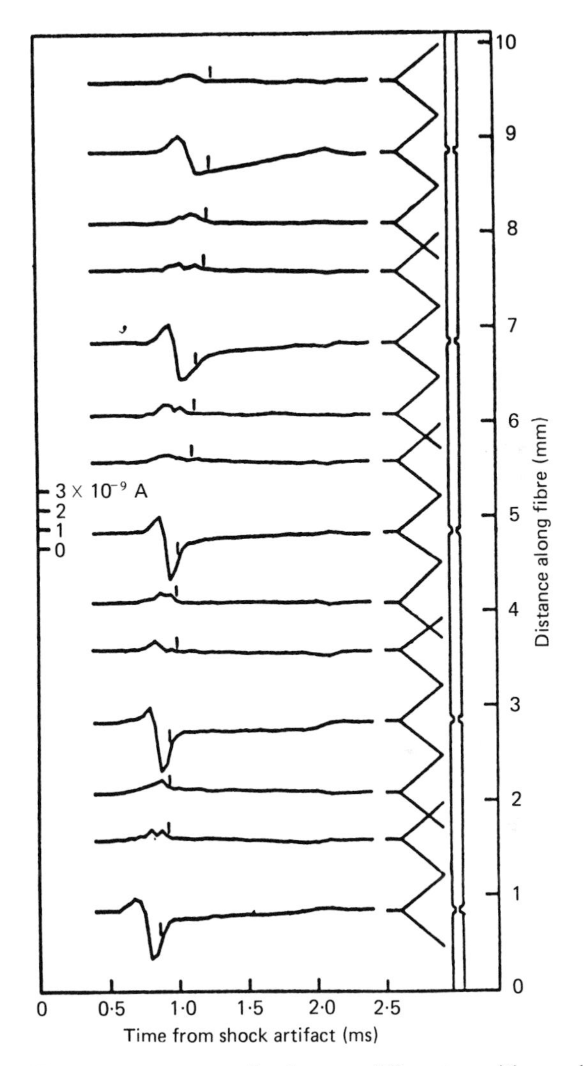

Fig. 6.5. Transverse currents flowing at different positions along an isolated frog nerve fibre. Each trace shows the difference between the longitudinal currents, recorded as in Fig. 6.4, at the two points 0.75 mm apart indicated to the right. The vertical mark above each trace shows the time when the change in membrane potential reached its peak at that position along the fibre. Outward current is plotted upwards. From Huxley and Stämpfli (1949).

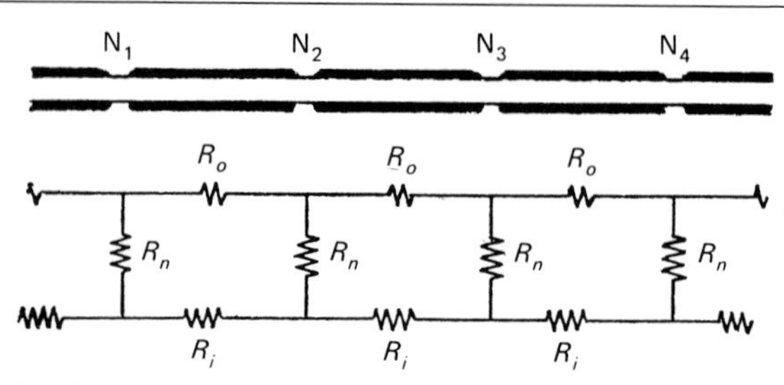

Fig. 6.6. Equivalent circuit for the resistive elements of a myelinated nerve fibre. According to Tasaki (1953), for a toad fibre whose outside diameter is 12 μm and a nodel spacing 2 mm, the internal longitudinal resistance R_I is just under 20 MΩ and the resistance R_n across each node is just over 20 MΩ. In a large volume of fluid the external resistance R_o is negligibly small.

potential is of the order of 120 mV, and since the threshold depolarization needed to excite the membrane is only about 15 mV, it again follows that the conduction mechanism works with an appreciable safety factor, and that the impulse should be able to encounter one or two inactive nodes without being blocked. Tasaki showed that two but not three nodes which had been treated with a local anaesthetic like cocaine could indeed be skipped.

A simple experiment which deserves mention was performed by Huxley and Stämpfli to demonstrate the importance of the external current pathway in propagation along a myelinated nerve fibre. The nerve of a frog's sciatic–gastrocnemius preparation was pared down until only one fibre was left (Fig. 6.7). Stimulation of the nerve at *P* then caused a visible contraction of a motor unit *M* in the muscle. The preparation was now laid in two pools of Ringer's solution on microscope slides *A* and *B* which were electrically insulated from one another, and its position was adjusted so that part of an internode, but not a node, lay across the 1 mm air gap separating the pools. At first, stimulation at *P* continued to cause a muscle twitch, but soon the layer of fluid outside the myelin sheath in the air gap was dried up by evaporation, and the muscle ceased to contract. Conduction across the gap could, however, be restored by placing a wet thread *T* between the two pools. This demonstrated that an action potential arriving at the node just to the left of the air gap could trigger the node on the far side of the gap only when there

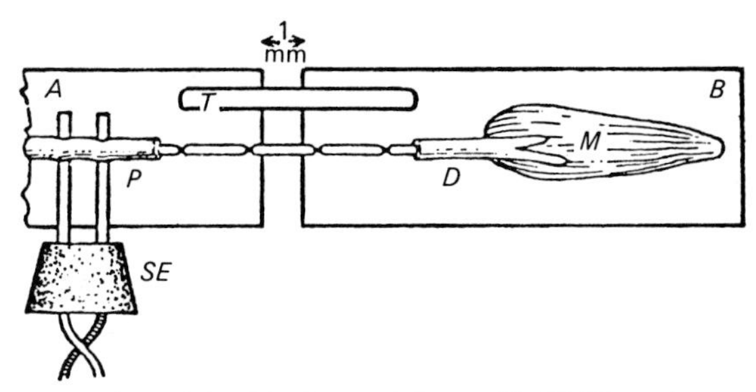

Fig. 6.7. Method used by Huxley and Stämpfli (1949) to demonstrate the role of the external current pathway in a myelinated nerve fibre. *A*, *B*, insulated microscope slides. *SE*, stimulating electrodes. *P*, proximal end of frog's sciatic nerve. *D*, distal end of nerve. *M*, gastrocnemius muscle. *T*, moist thread providing an electrical connection between the pools of Ringer's solution on the slides.

was an electrical connection between the pools whose resistance was fairly low. On reference to Fig. 6.6 it will be seen that if R_0 becomes at all large, the potential change at N_2 produced by a spike at N_1 will fall below the threshold for excitation. As has been pointed out by Tasaki, a reservation needs to be made about this experiment. Unless special precautions are taken, the stray electrical capacity between the pools may provide, for a brief pulse of current, an alternative pathway outside the dried-up myelin of the internode whose impedance may be low enough for excitation to occur at the further node if its threshold is low. Even with such precautions, Tasaki found that impulses were still able to jump the gap if the fibre had a really low threshold, probably because simple evaporation could not make the external resistance high enough. Nevertheless, the fact that the experiment works in a clear-cut way only if the threshold is somewhat higher than it is *in vivo* does not prevent it from proving rather satisfactorily that there must be a low-impedance pathway between neighbouring nodes *outside* the myelin sheath if the nerve impulse is to be conducted along the fibre.

FACTORS AFFECTING CONDUCTION VELOCITY

Since the passive electrotonic spread of potential along a nerve fibre is an almost instantaneous process, it may be asked why the nerve impulse is not propagated more rapidly than it actually is. In myelinated fibres the explanation is that there is a definite delay of about 0.1 ms at each node (see Fig. 6.4), which represents the time necessary for Na^+ ions to move through the membrane at the node in a quantity sufficient to discharge the membrane capacity and build up a reversed potential. Conduction in a non-myelinated fibre is slower than in a myelinated fibre of the same diameter because the membrane capacity per unit length is much greater, and the delay in reversing the potential across it arises everywhere and not just at the nodes. Because the time constant for an alteration of membrane potential depends both on the magnitude of the membrane capacity and on the amount of current that flows into it, conduction velocity is affected by the values of the resistances in the equivalent electrical circuit, and also by the closeness of packing of sodium channels in the membrane, which determines the sodium current density. The effects of changing R_o and R_i can best be seen in isolated axons. Thus Hodgkin (1939) showed that when R_o was increased by raising an axon out of a large volume of sea water into a layer of liquid paraffin, the conduction velocity fell by about 20% in a 30 μm crab nerve fibre and by 50% in a 500 μm squid axon; and when the axon was mounted in a moist chamber lying across a series of metal bars which could be connected together by a trough of mercury, the act of short-circuiting the bars increased the velocity by 20%. More recently, del Castillo and Moore (1959) showed that a reduction in R_i brought about by inserting a silver wire down the centre of a squid axon could greatly speed up conduction.

One of the reasons why large non-myelinated fibres conduct faster than small ones is the decrease of R_i with an increase of fibre diameter. Assuming the properties of the membrane to be identical for fibres of all sizes, it can be shown that conduction velocity should be proportional to the square root of diameter. Experimentally this does not always seem to hold good, a possible explanation being that one of the ways in which giant axons are specially adapted for rapid conduction is through an increase in the number of sodium channels in the membrane. Measurements of the binding of labelled TTX have shown that the smallest fibres of all, those in garfish olfactory nerve, have the fewest channels, the site density being 35 μm^{-2} as compared with 90 and 100 μm^{-2} in lobster leg nerve and rabbit vagus nerve respectively (Ritchie and Rogart, 1977). However, in the squid giant axon there are about 290 TTX binding sites μm^{-2} (Keynes and Ritchie, 1984). Since the flow of gating current has the consequence of increasing the effective size of the membrane capacity, there is an optimum sodium channel density above which the conduction velocity would fall off again. Hodgkin (1975) has calculated that the value found in squid is not far from the optimum.

FACTORS AFFECTING THE THRESHOLD FOR EXCITATION

As seen, for example, in Fig. 2.7, excitation of a nerve fibre involves the rapid depolarization of the membrane to a critical level normally about 15 mV less negative than the resting potential. The critical level for excitation is the membrane potential at which the net rate of entry of Na^+ ions becomes exactly equal to the net rate of exit of K^+ ions plus a small contribution from an entry of Cl^- ions. Greater depolarization than this tips the balance in favour of Na^+, and the regenerative process described in Chapter 4 takes over and causes a rapidly accelerating inrush of sodium. After just subthreshold depolarization, when g_{Na} will have been raised over an appreciable area of membrane, the return to the resting potential will be somewhat slow at first, and a non-propagated *local response* may be observed.

At the end of the spike the membrane is left with its sodium permeability mechanism inactivated and its potassium permeability appreciably greater than normal. Both changes tend to raise the threshold for re-excitation. The partial inactivation of the sodium permeability system means that even to raise inward Na^+ current to the normal critical value requires more depolarization than usual, and the raised potassium permeability means that the critical Na^+ current is actually above normal. Until the permeabilities for both ions have returned to their resting levels, and the sodium permeability system is fully reactivated, the shock necessary to trigger a second spike is above the normal threshold in size.

It has long been known that nerves are not readily stimulated by slowly rising currents, because they tend to *accommodate* to this type of stimulus. Accommodation arises partly because sustained depolarization brings about a long-lasting rise in potassium permeability, and partly because at the same time it semi-permanently inactivates the sodium permeability mechanism. Both changes take place with an appreciable lag after the membrane potential is lowered, so that they are not effective when a constant current is first applied, but become important after a little while. They also persist for some time after the end of a stimulus, and so are responsible for the appearance of *post-cathodal depression*, which is a lowering of excitability after prolonged application of a weak cathodal current. As a result of accommodation, cathodal currents that rise more slowly than a certain limiting value do not stimulate at all, since the rise in threshold keeps pace with the depolarization.

Another familiar phenomenon is the occurrence of excitation when an anodal current is switched off. This *anode break excitation* can readily be demonstrated in isolated squid axons or frog nerves, but is not seen in freshly dissected frog muscle or in nerves stimulated *in situ* in living animals. The conditions under which anode break excitation is exhibited are that the resting potential should be well below E_K because of a steady leakage of potassium. The nerve can then be considered to be in a state of mild cathodal depression, with g_{Na} partially inactivated and g_K well above normal. The effect of anodal polarization of the membrane is to reactivate the sodium permeability system and to reduce the potassium permeability, and this improved state persists for a short while after the current is switched off. While it lasts, the critical potential at which inward Na^+ current exceeds outward K^+ current may be temporarily above the membrane potential in the absence of external current. When the current is turned off, an action potential is therefore initiated.

Divalent ions like Ca^{2+} and Mg^{2+} strongly affect the threshold behaviour of excitable membranes. In squid axons, even a slight reduction in external $[Ca^{2+}]$ may set up a sinusoidal oscillation of the membrane potential, while a more drastic reduction of the calcium will result in a spontaneous discharge of impulses at a high repetition frequency. Conversely, a rise in external [Ca^{2+}] helps to stabilize the membrane and tends to raise the threshold for excitation. Changes in external [Mg^{2+}] have rather similar effects on peripheral nerves, magnesium being about half as effective as calcium in its stabilizing influence. Voltage-clamp studies by Frankenhaeuser and Hodgkin (1957) have shown that the curve relating peak sodium conductance to membrane potential is shifted in a positive direction along the voltage axis by raising [Ca^{2+}], and is shifted in the opposite direction by lowering [Ca^{2+}]. However, the resting potential is rather insensitive to changes in [Ca^{2+}]. This readily explains the relationship between [Ca^{2+}] and threshold, since a rise in [Ca^{2+}] moves the critical triggering level away from the resting potential, while in a fall in [Ca^{2+}] moves the critical level towards it. A moderate reduction in [Ca^{2+}] may bring the critical level so close to the resting potential that the membrane behaves in an unstable and oscillatory fashion, and a further reduction will then increase the amplitude of the oscillations to the point where they cause a spontaneous discharge of spikes. Although calcium and magnesium have similar actions on the excitability of nerve and muscle fibres, they have antagonistic actions at the neuromuscular junction (see p. 104) and at some synaptic junctions between neurons, because calcium increases the amount of acetylcholine released by a motor nerve ending, and magnesium reduces it. Changes in the plasma calcium level in living animals may therefore give rise to tetany, but there is a complicated balance between central and peripheral effects.

AFTER-POTENTIALS

In many types of nerve and muscle fibre the membrane potential does not return immediately to the base-line at the foot of the action potential, but undergoes further slow variations known as after-potentials. The nomenclature of after-potentials dates from the period before the invention of intracellular recording techniques when external electrodes were used, so that an alteration of potential in the same direction as the spike itself is termed a *negative after-potential*, while a variation in the opposite direction corresponding to a hyperpolarization of the membrane is termed a *positive after-potential* (see Fig. 2.3). As may be seen in Fig. 2.4*b*, isolated squid axons display a characteristic *positive phase* which is almost completely absent in the living animals (Fig. 2.4*a*), while frog muscle fibres have a prolonged negative after-potential (Fig. 2.4*h*). In some mammalian nerves, both myelinated and non-myelinated, there is first a negative and then a positive after-potential. A related phenomenon, which is most marked in the smallest fibres, is the occurrence after a period of repetitive activity of a prolonged hyperpolarization of the membrane known as the post-tetanic hyperpolarization.

There is no doubt that after-potentials are always connected with changes in membrane permeability towards specific ions, but there is more than one way in which the membrane potential can be displaced either upwards or downwards. In the isolated squid axon, for example, the positive phase arises because the potassium conductance is still relatively high at the end of the spike, whereas the sodium conductance is inactivated and is therefore below normal. The membrane potential consequently comes close to E_K for a short while, and then drops back as g_K and g_{Na} resume their usual resting values. The mechanism responsible for production of the positive after-potential and the post-tetanic hyperpolarization in vertebrate nerves is quite different, for it has been shown to involve an enhanced rate of extrusion of Na^+ ions by the sodium pump operating in an electrogenic mode. In other cases, a change in the relative permeability of the membrane to Cl^- and K^+ ions may play a part. There is also evidence that the presence of Schwann cells partially or wholly enveloping certain types of nerve fibre has important effects on the after-potential by slightly restricting the rate of diffusion of ions in the immediate neighbourhood of the nerve membrane.

CHAPTER SEVEN

Neuromuscular transmission

Skeletal muscles are innervated by motor nerves. Excitation of the motor nerve is followed by excitation and contraction of the muscle. Thus excitation of one cell, the nerve axon, produces excitation of another cell which it contacts, the muscle fibre. The region of contact between the two cells is called the *neuromuscular junction*. The process of the transmission of excitation from the nerve cell to the muscle cell is called *neuromuscular transmission*. This chapter is concerned with how this process occurs.

Regions at which transfer of electrical information between a nerve cell and another cell (which may or may not be another nerve cell) occurs are known as *synapses*, and the process of information transfer is called *synaptic transmission*. Neuromuscular transmission is just one form of synaptic transmission; we shall examine the properties of some other synapses in the following chapter.

THE NEUROMUSCULAR JUNCTION

Each motor axon branches so as to supply an appreciable number of muscle fibres. Fig. 7.1 shows the arrangement in most of the muscle fibres in the frog. Each axon branch loses its myelin sheath where it contacts the muscle cell and splits up into a number of fine terminals which run for a short distance along its surface. The region of the muscle fibre with which the terminals make contact is known as the *end-plate*. Structures and events occurring in the axon are called *presynaptic* whereas those occurring in the muscle cell are called *postsynaptic*.

Further details of the structure in the junctional region can be determined by electron microscopy of thin sections, with the results shown in Fig. 7.2. The nerve cell is separated from the muscle cell by a gap about 50 nm wide between the two apposing cell

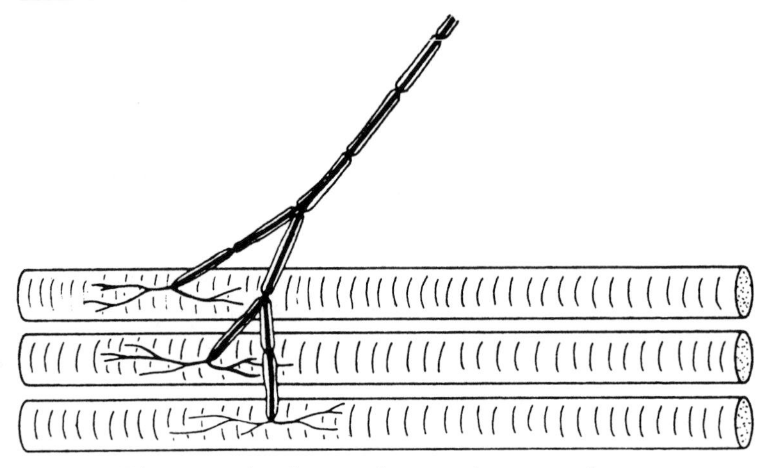

Fig. 7.1. Diagrammatic picture of a vertebrate muscle motor nerve terminal. In most cases a single motor axon innervates many more muscle fibres than the three shown here.

membranes. This gap is called the *synaptic cleft* and is in contact at its edges with the other extracellular spaces of the body. The axon terminal contains a large number of small membrane-bound spheres, the *synaptic vesicles*, and also a considerable number of mitochondria. We shall see that the synaptic vesicles and the synaptic cleft play a vital role in the transmission process at this and other synapses.

The muscle cell membrane immediately under the axon terminal is thrown into a series of folds. The outside of the axon terminal is covered with a Schwann cell and the whole is held in position by some collagen fibres.

CHEMICAL TRANSMISSION

We must now consider the question, how does excitation in the presynaptic cell produce a response in the postsynaptic cell? One possibility, first suggested in the nineteenth century, is that the presynaptic cell might release a chemical substance which would then act as a messenger between the two cells.

An experiment to test this idea for the frog heart was carried out

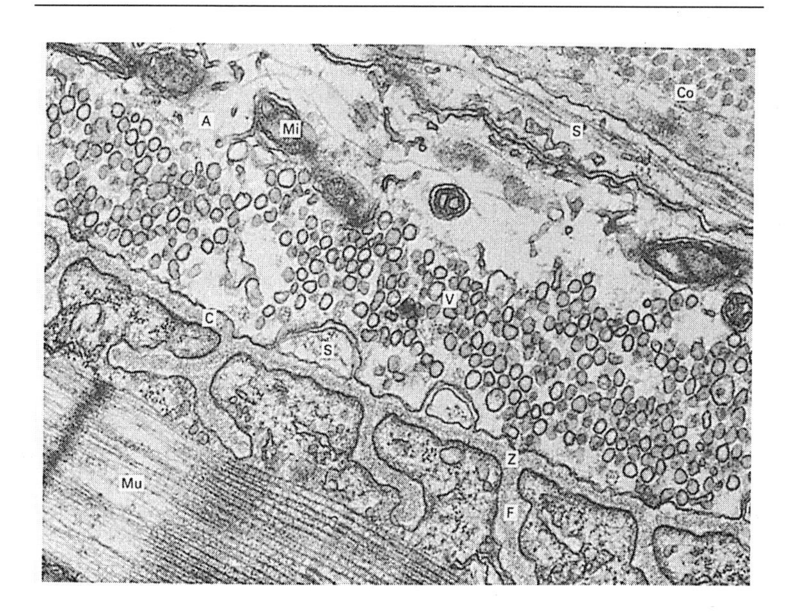

Fig. 7.2. Electron micrograph showing the structure of the frog neuromuscular junction. The axon terminal (A) runs diagonally across the middle of the section, covered by a Schwann cell (S) and collagen fibres (Co), and overlying a muscle cell (Mu). Between the axon and the muscle cell is the synaptic cleft (C). The acetylcholine receptors are concentrated at the top of a series of folds (F) in the subsynaptic membrane. The terminal contains mitochondria (Mi) and large numbers of synaptic vesicles (V). Vesicle release probably occurs at presynaptic active zones (Z). Magnification 27000 times. Photograph supplied by Professor J. E. Heuser.

by O. Loewi in 1921. The heart normally beats spontaneously, but it can be inhibited by stimulation of the vagus nerve. Loewi found that the perfusion fluid from a heart which was inhibited by stimulation of the vagus would itself reduce the amplitude of the normal beat in the absence of vagal stimulation. Perfusion fluid from a heart beating normally did not have this effect. This means that stimulation of the vagus results in the release of a chemical substance, presumably from the nerve endings.

It did not take too long to show that the chemical substance was acetylcholine (Fig. 8.1), a substance whose pharmacological action had previously been demonstrated by H. H. Dale. Dale and his colleagues then went on to show that acetylcholine was released from the motor nerve endings in skeletal muscle when the motor nerves were stimulated electrically.

It seems then that synaptic transmission is in most cases a chemically mediated process involving the release of a *transmitter substance* from the presynaptic terminals. As we shall see later, acetylcholine is an important transmitter substance, but it is not the only one.

POSTSYNAPTIC RESPONSES

The end-plate potential

Stimulation of the motor nerve produces electrical responses in the muscle fibre. Our understanding of the nature of these events was greatly increased when P. Fatt and B. Katz used intracellular electrodes to the study the problem in 1951. Fig. 7.3 shows the experimental arrangement which they used. The glass micropipette electrode, filled with a concentrated solution of potassium chloride, is inserted into the muscle fibre in the end-plate region. A suitable amplifier then measures the voltage between the tip of that electrode and another electrode in the external solution, so giving the electrical potential difference across the membrane. The muscle may be treated with curare (an arrow poison used by South American Indians, which causes paralysis) so as to partly block the

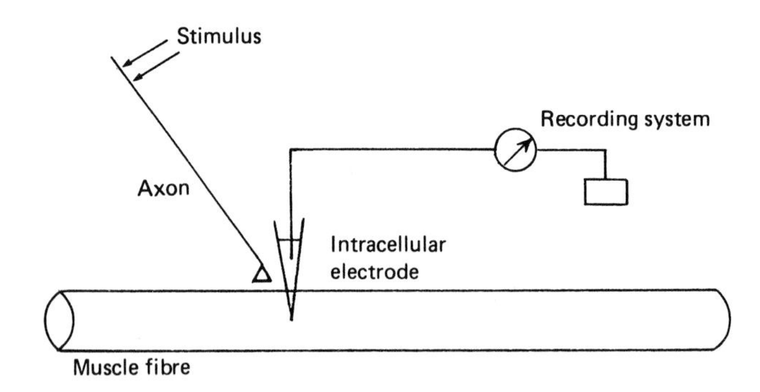

Fig. 7.3. Diagram to show how the end-plate potential is recorded from a frog muscle fibre using an intracellular microelectrode.

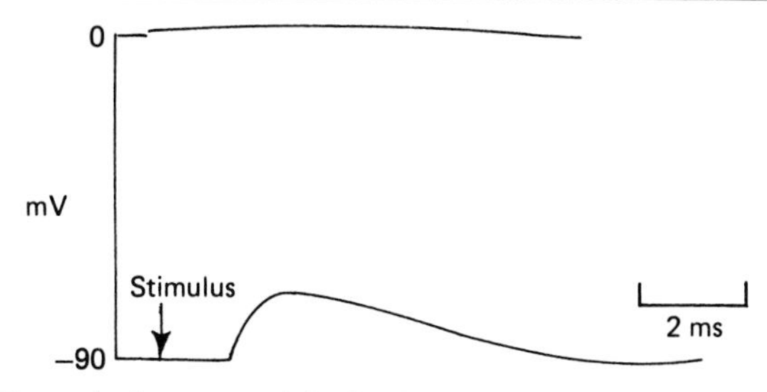

Fig. 7.4. The end-plate potential of a frog muscle fibre in the presence of curare.

transmission process. The nerve can be stimulated via a couple of silver wire electrodes.

Fig. 7.4 shows the response seen in the presence of a moderate amount of curare. There is a rapid depolarization of a few millivolts, followed by a rather slower return to the resting membrane potential. This response is not seen if the microelectrode is inserted at some distance from the end-plate, and hence it is called the *end-plate potential.* Increasing the curare concentration reduces the size of the end-plate potential

Excitation of the muscle fibre

In the absence of curare the end-plate potential is its normal size, and the response recorded at the end plate is more complicated in form, as is shown in Fig. 7.5*a*. The muscle cell membrane is electrically excitable so that it will carry all-or-nothing propagated action potentials just as in the nerve axon. In the absence of curare the end-plate potential is large enough to cross the threshold for electrical excitation of the muscle cell membrane, so that an action potential arises from it and propagates along the length of the muscle fibre. The record in Fig. 7.5*a* is thus a combination of end-plate potential and action potential. At some distance from the end-plate, a 'pure' action potential, free from the complicating effects of the end-plate potential, can be seen, as in Fig. 7.5*b*.

The ionic basis of the muscle action potential is much the same as in the nerve axon. It is reduced in size or blocked in low sodium ion concentrations or in the presence of tetrodotoxin. So we can assume

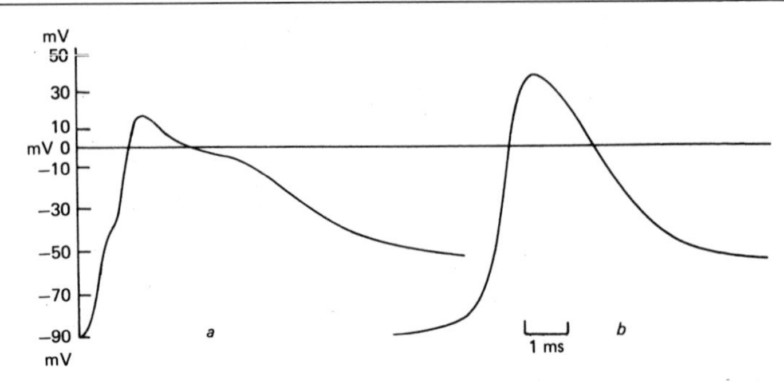

Fig. 7.5. Action potentials produced in a frog muscle fibre by stimulation of its motor nerve axon. In trace *a* the microelectrode was positioned at the end-plate region, so that the response includes an end-plate potential plus the action potential which it gives rise to. In trace *b* the microelectrode was positioned at some distance from the end-plate so that no end-plate potential component is recorded. Based on Fatt and Katz (1951).

that the cell membrane contains separate channels for sodium and potassium ions, both types being opened by a suitable change in membrane potential. The muscle action potential triggers the contraction of the muscle, as we shall see in Chapter 10.

The response to acetylcholine

We have seen that stimulation of the motor nerve causes the release of acetylcholine. Is the end-plate potential a direct result of the action of this acetylcholine on the postsynaptic membrane? Clearly a good way to test this idea is to apply acetylcholine to the postsynaptic membrane and see if it produces a depolarization. If it does not, we shall have to think again, but if it does, then the idea becomes much better established as a result of passing this test.

The best way of applying acetylcholine to the postsynaptic membrane is by means of a technique known as ionophoresis or iontophoresis. Acetylcholine is a positively charged ion (Fig. 8.1), and so it will move in an electric field. Hence it can be ejected from a small pipette by passing current through the pipette, as is shown in Fig. 7.6. Thus a brief, highly localised 'pulse' of acetylcholine can be applied to the postsynaptic membrane. Such a pulse produces a depolarization of the muscle fibre which is very similar to the end-

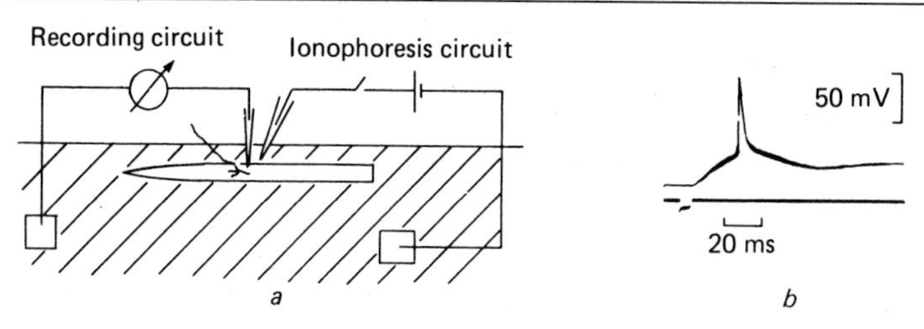

Fig. 7.6. The ionophoresis technique applied to a frog muscle fibre (*a*), and the response to a pulse of acetylcholine applied by this method (*b*). From del Castillo and Katz (1955).

plate potential (Fig. 7.6*b*). The pharmacological properties of the response are also similar to those of the end-plate potential; it is reduced in the presence of curare, for example. So it seems very reasonable to conclude that the end-plate potential is indeed produced by the acetylcholine which is released from the motor nerve ending.

Ionic current flow during the end-plate potential

Fatt and Katz suggested that the end-plate potential was produced by a general increase in the ionic permeability of the postsynaptic membrane. A closer look at the problem was provided by A. and N. Takeuchi, who used a voltage-clamp technique on frog muscle fibres to examine the postsynaptic current flow during the response to a nerve impulse. They found that the duration of the end-plate current flow is briefer than the end-plate potential, as is shown in Fig. 7.7. This is because the 'tail' of the end-plate potential is caused by a recharging of the membrane capacitance, which does not occur when the membrane potential is clamped.

The membrane potential can be clamped at different values. When this is done it is found that the amplitude of the end-plate current varies linearly with membrane potential (Fig. 7.8). This linear relation between current and voltage is just what we would expect from Ohm's law: it indicates that the conductance of the membrane at the peak of the end-plate current is constant and not affected by the membrane potential. This is in marked contrast to the situation in the nerve axon, where the sodium and potassium

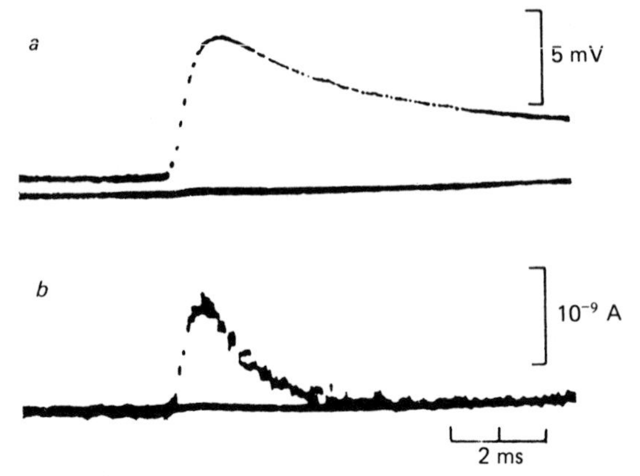

Fig. 7.7. Measurement of the end-plate potential and end-plate current in a curarized frog muscle fibre. Trace *a* shows the EPP, recorded in the usual manner. Trace *b* shows the EPC recorded from the same end-plate with the membrane potential clamped at its resting level. From Takeuchi and Takeuchi (1959).

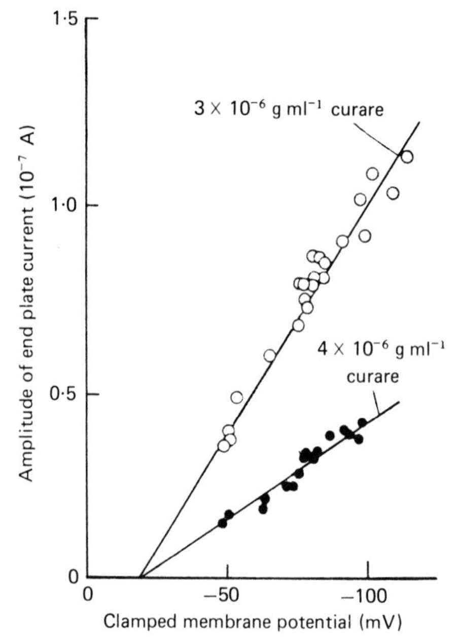

Fig. 7.8. Results of a voltage clamp experiment on a curarized frog muscle fibre, showing that the peak end-plate current varies linearly with membrane potential. From Takeuchi and Takeuchi (1960).

permeabilities of the membrane are most strongly altered by changes in membrane potential.

The point at which the line through the experimental points in Fig. 7.8 crosses the voltage axis is called the *reversal potential*. If the membrane potential is made more positive than this we would expect the end-plate current to flow in the opposite direction, as indeed it does.

The reversal potential is the membrane potential at which the net ionic current is zero. If only one type of ion were flowing, the reversal potential would be at the equilibrium potential for that ion; for example, it would be about +50 mV if only sodium ions were flowing. But if more than one type of ion flows during the end-plate current, then the reversal potential will be somewhere between the various equilibrium potentials for the different ions. The actual reversal potential, at about −15 mV in the Takeuchis' experiments, is compatible with the idea that both sodium and potassium ions flow during the end-plate current. If we alter the equilibrium potential for one of the ions involved, then the reversal potential will also alter. The Takeuchis did just this by changing the ionic concentrations in the external solutions. They found that alterations in sodium and potassium ion concentrations both altered the reversal potential, whereas alterations in chloride ion concentration did not. This means that the end-plate current consists of a flow of sodium and potassium ions. To put it another way, acetylcholine increases the permeability of the postsynaptic membrane to both sodium and potassium ions simultaneously.

Acetylcholine receptors and single channel responses

It is now generally accepted that most pharmacological agents act at specific molecular sites on the cell membrane, called *receptors*. Only cells which possess the appropriate receptor will respond to a particular agent. In accordance with this view, we would expect to find specific acetylcholine receptors on the postsynaptic membrane at the neuromuscular junction.

The substance α-bungarotoxin, a polypeptide found in the venom of a Formosan snake, causes neuromuscular block by binding tightly to the acetylcholine receptors. Using radioactive toxin (made by acetylating the toxin with ^{3}H-acetic anhydride) it is a simple matter to show by autoradiography that the toxin rapidly becomes attached to the postsynaptic membrane at the end-plate regions. By counting the grains of silver produced in the autoradiograph, it is then possible to count the number of toxin molecules that have been bound and from this to estimate the number of receptors present. The results suggest that there are about 3×10^7 binding sites per end-plate in mammals, corresponding to an average density in the region of 10^4 sites per μm^2.

Since combination of acetylcholine with the receptors causes an increase in membrane permeability to sodium and potassium ions, it seems very likely that each receptor is closely associated with a *channel* through which this ionic flow can occur. It would normally be closed and would open for a short time when acetylcholine combines with the receptor.

Direct evidence for this view was provided in experiments by E. Neher and B. Sakmann, for which they developed the patch-clamp technique (see Fig. 4.17). They used frog muscle fibres whose motor nerve supply had been cut some time previously. Following this procedure the whole surface of the fibre becomes sensitive to acetylcholine and contains a low density of acetylcholine receptors.

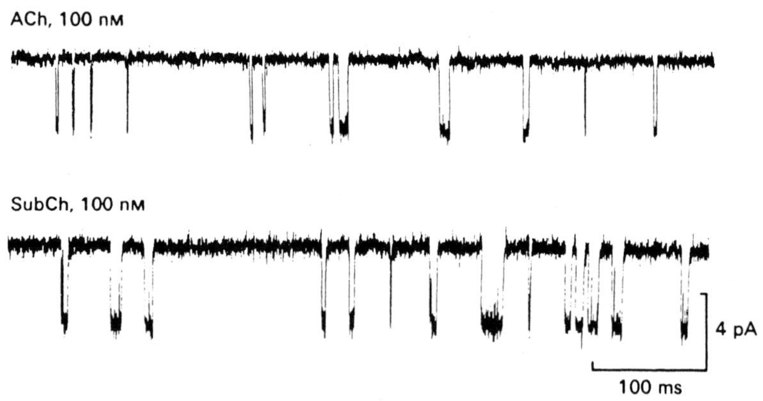

Fig. 7.9. Patch-clamp records of single channel currents from frog end-plates. The upper records shows the response to acetylcholine, the lower one shows the response to suberyldicholine. From Colquhoun and Sakmann (1985).

The polished tip of a microelectrode can then be pushed against the fibre membrane, so that the current flow through a small patch of membrane containing only a few receptors can be measured.

Neher and Sakmann found that, when the patch electrode contains acetylcholine, individual channels each produce a square pulse of current lasting up to a few milliseconds. The durations of successive current pulses were variable, but their amplitudes were constant, as is shown in Fig. 7.9. This suggests that the channel is either open or shut, and that it can only open when it combines with acetylcholine. Probably two molecules of acetylcholine need to combine with each receptor in order to open the channel. The end-plate potential is thus produced when a large number of channels open more or less at the same time.

The molecular structure of the acetylcholine receptor

The electric organs of the electric ray *Torpedo* provide a rich source of acetylcholine receptors. They can be isolated by using their specific binding to the snake venom α-bungarotoxin. The receptors are pentameric proteins with a total molecular weight of about 290 kDa. The subunits are called the α, β, γ and δ chains; there are two α chains in each receptor and one of each of the others. The binding sites for acetylcholine are located on the α chains.

The acetylcholine receptor was the first ionic channel to be sequenced by using recombinant DNA techniques. In 1982 the Kyoto University group (Noda and his colleagues) published the amino acid sequence of the α subunit, and sequences for the other subunits soon followed. The subunits varied in size from 437 amino acids (50 kDa) for the α chain to 501 amino acids (58 kDa) for the δ chain.

The sequences for the four subunits show considerable homology. They all have four hydrophobic segments which probably form membrane-crossing helices (Fig. 7.10). The long section from the beginning of the chain to the first membrane-crossing helix is apparently all on the outside of the membrane. It contains disulphide crosslinks and sites for the attachment of sugars. In the α chain it contains the sites for binding acetylcholine and α-bungarotoxin.

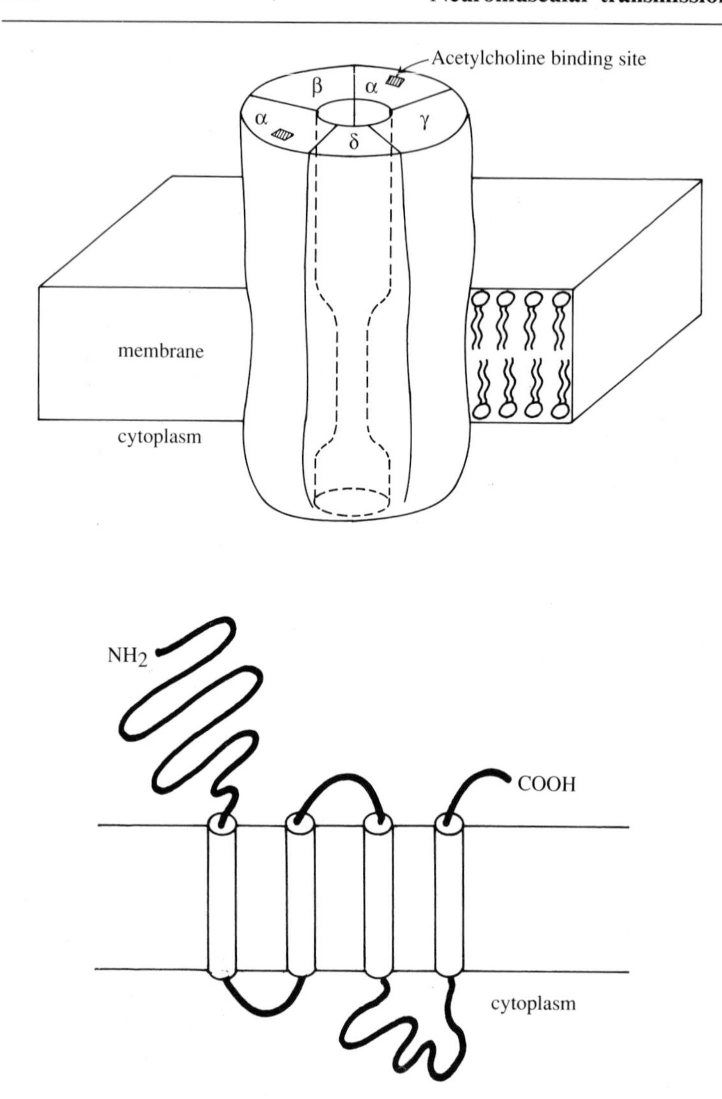

Fig. 7.10. Diagrams of the molecular structure of the nicotinic acetylcholine receptor/channel. The complex consists of five subunits (upper diagram); the two α subunits contain acetylcholine binding sites. The amino acid chain of each subunit contains four membrane-crossing segments (lower diagram). The diagrams should not be taken too literally.

How can we be sure that the four subunits are sufficient to produce a functional receptor? The oocytes of the African clawed toad *Xenopus* have provided a most useful method for solving this problem.

Oocytes are large cells which are about to develop into mature eggs. They possess the normal translation machinery and so they will respond to the injection of messenger RNA by making the protein for which it codes. Barnard and his colleagues injected oocytes with messenger RNA from *Torpedo* electric organ; two days later the oocyte would respond to application of acetylcholine by ionophoresis with a rapid depolarization, as is shown in Fig. 7.11.

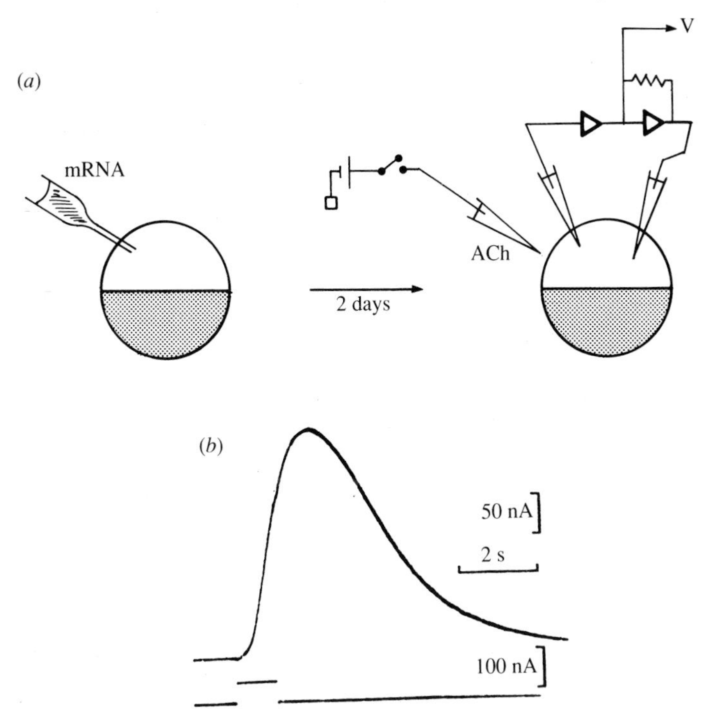

Fig. 7.11. Expression of acetylcholine receptors in *Xenopus* oocytes. Messenger RNA from *Torpedo* electric organ is injected into an oocyte; two days later the oocyte membrane potential is voltage-clamped while acetylcholine is applied by ionophoresis (*a*). The record (*b*) shows the current response (upper trace) to acetylcholine; the lower trace monitors the ionophoresis current. From Barnard, Miledi and Sumikawa (1982).

Molecular cloning methods can be used to make messenger RNA coding for the different subunits of the acetylcholine receptor. Only when messenger RNAs for all four of the subunits were injected would the oocyte respond to application of acetylcholine. This shows that all four of the subunits are necessary for production of a functional receptor. It also shows that no extra components are required, so providing excellent confirmation for the conclusions of the recombinant DNA work.

Substances other than acetylcholine can combine with the receptors. Blocking agents such as α-bungarotoxin and curare combine with the receptors without opening their channels. Curare and some other compounds of this type are useful as muscular relaxing agents in surgery. Agonists of acetylcholine, such as nicotine and carbachol, combine with the receptors and do open the channels, so they induce ionic flow just as acetylcholine does.

In the disease myasthenia gravis it seems likely that the body produces antibodies to the neuromuscular acetylcholine receptor, resulting in partial neuromuscular block.

Acetylcholinesterase

The enzyme acetylcholinesterase hydrolyses acetylcholine to form choline and acetic acid. Histochemical staining shows that this enzyme is greatly concentrated in the synaptic cleft and especially in the folds of the subsynaptic membrane. Its function is to hydrolyse the acetylcholine so as to limit the time during which it is active after being released by a motor nerve impulse.

A number of substances inhibit the action of acetylcholinesterase; they are known as *anticholinesterases*. They include the alkaloid eserine and the organophosphorus insecticides.

PRESYNAPTIC EVENTS

The presynaptic nerve terminal is much smaller than the postsynaptic muscle fibre, and so it is much more difficult to investigate its properties directly. It is not possible, for example, to insert an intracellular microelectrode into it. Fortunately, however, the main feature of interest to us is the release of acetylcholine, and we can measure this relatively easily by recording the responses of the postsynaptic cell.

The quantal release of acetylcholine

In the resting muscle small fluctuations in membrane potential occur at the end-plate region (Fig. 7.12). They follow much the same time course as do end-plate potentials, but are only about 0.5 mV in size, hence they are called *miniature end-plate potentials*. They are reduced in size by curare and increased in size by anti-cholinesterases, and so it looks as though they are produced by the spontaneous release of 'packets' of acetylcholine from the motor nerve ending. Kuffler and Yoshikami compared their size with those of responses to ionophoretic application of acetylcholine; they concluded that each miniature end-plate potential is produced by the action of just under 10000 molecules of acetylcholine.

An excess of magnesium ions blocks neuromuscular transmission by reducing the amount of acetylcholine released per nerve impulse. Del Castillo and Katz found that the size of the small end-plate potentials produced under these conditions fluctuates in a stepwise manner. Each step was about the size of a miniature end-plate potential. They therefore suggested that acetylcholine is released from the motor nerve terminals in discrete 'packets' or *quanta*. The

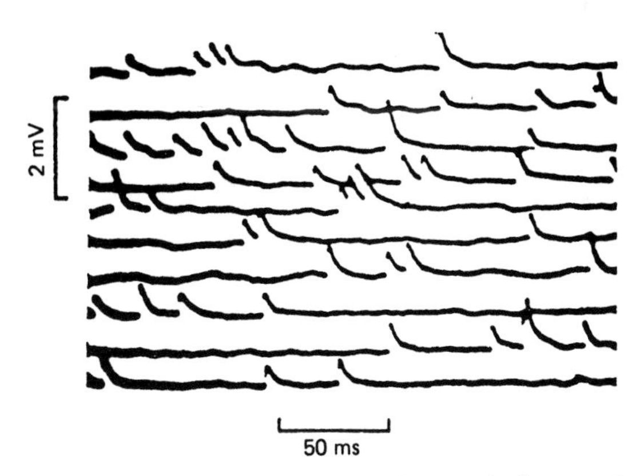

Fig. 7.12. A series of membrane potential records from a frog neuromuscular junction showing miniature end-plate potentials. From Fatt and Katz (1952).

normal end-plate potential is then the response to some hundreds of these quanta, all released at the same time following the arrival of a nerve impulse at the axon terminal. Miniature end-plate potentials are the result of spontaneous release of single quanta.

But why should acetylcholine be discharged from nerve endings in packets of nearly 10000 molecules? The axon terminals contain large numbers of *synaptic vesicles* about 50 nm in diameter (Fig. 7.2). Similar vesicles have been found in the presynaptic terminals at other synapses where chemical transmission occurs. Del Castillo and Katz suggested that they contain the chemical transmitter substance (acetylcholine at the neuromuscular junction), and that the discharge of the contents of one vesicle into the synaptic cleft corresponds to the release of one quantum of the transmitter.

Confirmation of this idea has since been provided by biochemical separation techniques, especially by Whittaker and his colleagues. A tissue containing a large number of nerve endings (such as brain tissue or electric organ) is first homogenized and then centrifuged. Nearly pure fractions of synaptic vesicles have been obtained from electric organs by this method, and it is found that they contain acetylcholine. They also contain some adenosine triphosphate (ATP), but the function of this is not too clear.

Depolarization and calcium entry

The normal trigger for the release of acetylcholine is the arrival of a nerve action potential at the terminal. Katz and Miledi found that, if the action potential is blocked by tetrodotoxin, acetylcholine can still be release by depolarizing the terminal with applied current. However, this current is only effective when calcium ions are present in the external solution. Axon terminals contain numbers of voltage-gated calcium channels. It seems probable that these open during the action potential so that calcium ions enter the terminal and somehow act as the trigger for the release of the contents of the vesicles.

The release-promoting effects of calcium are antagonized by magnesium ions. Perhaps the two ions compete for some site to form an intermediate compound which is necessary for the release process, but which is ineffective when it is in the form of a magnesium complex.

Synaptic delay

There is a short delay between the arrival of an action potential in the terminal and the ensuing depolarization of the muscle cell. In frog muscle at 17 °C, the minimum time is about 0.5 ms. The major part of this delay is probably taken up by the processes involving calcium ions in the presynaptic terminal. In addition the acetylcholine takes a little time to diffuse across the synaptic cleft and to combine with the acetylcholine receptors to open the ionic channels.

Facilitation and depression

When the motor axon is stimulated repetitively, the successive end-plate potentials produced are often of different sizes. At the beginning of a series, and especially in low calcium ion concentrations, successive potentials increase in size. This phenomenon is known as *facilitation*. The larger responses are made up of more quantal units. Thus facilitation seems to be caused by an enhancement of the release process, perhaps because of an accumulation of calcium ions at some site within the presynaptic terminal.

The opposite of facilitation is *depression*, in which succeeding responses are smaller and composed of fewer quantal units. Depression occurs when a large number of quanta have recently been released, as in the later stages of a train of stimuli in the presence of an adequate concentration of calcium ions. It appears to be caused by a temporary reduction in the number of vesicles which are available for release.

CHAPTER EIGHT

Synaptic transmission in the nervous system

The functioning of the nervous system depends largely on the interactions between its constituent nerve cells, and these interactions take place at synapses. In most cases synaptic transmission is chemical in nature, so that, as in neuromuscular transmission, the presynaptic cell releases a chemical transmitter substance which produces a response in the postsynaptic cell. There are a few examples of electrically transmitting synapses, which we shall consider briefly at the end of this chapter.

Acetylcholine is only one of a range of different neurotransmitters. Fig. 8.1 shows some of the variety found in the central nervous system. For a long time it was thought that any one cell would only release one neurotransmitter, but several cases where two of them are released at the same time are now known.

Different chemically transmitting synapses differ in the details of their anatomy, but some features are common to all of them. In the presynaptic terminal the transmitter substance is packaged in synaptic vesicles. The pre- and postsynaptic cells are separated by a synaptic cleft into which the contents of the vesicles are discharged. There are specific receptors for the neurotransmitter on the postsynaptic membrane.

Just as with the neuromuscular junction, our knowledge of how synapses work was greatly affected by the invention of the intracellular microelectrode. Much of the fundamental work with this technique was performed by J. C. Eccles and his colleagues on the spinal motoneurons of the cat, so it is with these that we shall begin our account of synapses between neurons.

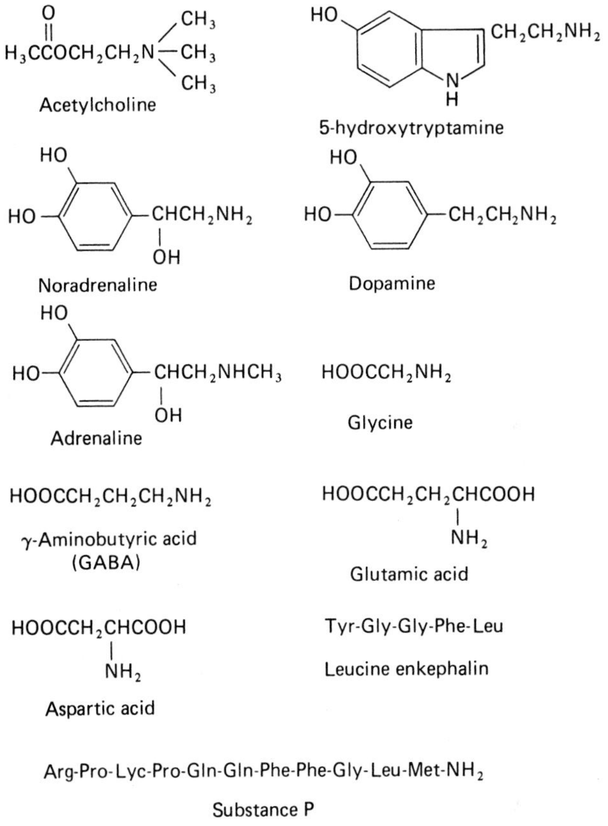

$H_3CCOCH_2CH_2N(CH_3)_3$ (with C=O)

Acetylcholine

HO– (indole ring) –$CH_2CH_2NH_2$, NH

5-hydroxytryptamine

HO, HO– (benzene ring) –$CHCH_2NH_2$ | OH

Noradrenaline

HO, HO– (benzene ring) –$CH_2CH_2NH_2$

Dopamine

HO, HO– (benzene ring) –$CHCH_2NHCH_3$ | OH

Adrenaline

$HOOCCH_2NH_2$

Glycine

$HOOCCH_2CH_2CH_2NH_2$

γ-Aminobutyric acid
(GABA)

$HOOCCH_2CH_2CHCOOH$ | NH_2

Glutamic acid

$HOOCCH_2CHCOOH$ | NH_2

Aspartic acid

Tyr-Gly-Gly-Phe-Leu

Leucine enkephalin

Arg-Pro-Lyc-Pro-Gln-Gln-Phe-Phe-Gly-Leu-Met-NH_2

Substance P

Fig. 8.1. Some transmitter substances in the central nervous system. From Ryall (1979).

SYNAPTIC EXCITATION IN MOTONEURONS

Motoneurons are the nerve cells which directly innervate skeletal muscle fibres. Their cell bodies lie in the ventral horn of the spinal cord, and their axons pass out to the peripheral nerves via the ventral roots. The cell body, or soma, is about 70 μm across, and extends into a number of fine branching processes, the dendrites, which may be up to 1 mm long. The surface of the soma and dendrites is covered with small presynaptic nerve terminals, and these regions of contact show the typical features of chemically transmitting synapses: a synaptic cleft and synaptic vesicles in the presynaptic cell.

Intracellular recording shows that motoneurons have a resting potential of about −70 mV. Depolarization of the membrane by about 10 mV results in the production of an action potential which propagates along the axon to the nerve terminals. Experiments involving the injection of various ions into motoneurons indicate that the ionic basis of their resting and action potentials is much the same as in squid axons. That is to say, the resting potential is slightly less than the potassium equilibrium potential, the action potential is caused primarily by a regenerative increase in sodium permeability, and the ionic gradients necessary for these potentials are dependent upon an active extrusion of sodium ions.

Excitatory postsynaptic potentials

Some of the presynaptic terminals on any particular motoneuron are the endings of sensory axons (known as group Ia fibres) from muscle spindles in the muscle which the motoneuron innervates. Stretching the muscle excites these axons, which may then excite the motoneurons supplying the muscle so that it contracts. This system is known as a monosynaptic reflex (Fig. 8.2). The knee-jerk reflex is a familiar example.

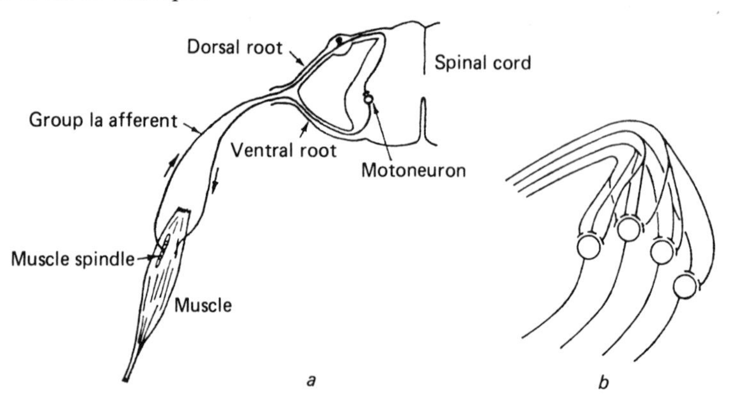

Fig. 8.2. Anatomical organization of the monosynaptic stretch reflex system (*a*). This diagram is much simplified: there are in fact very many stretch receptors and afferent and efferent neurons associated with each muscle. Diagram (*b*) indicates how the afferent fibres branch to synapse with different members of the motoneuronal pool.

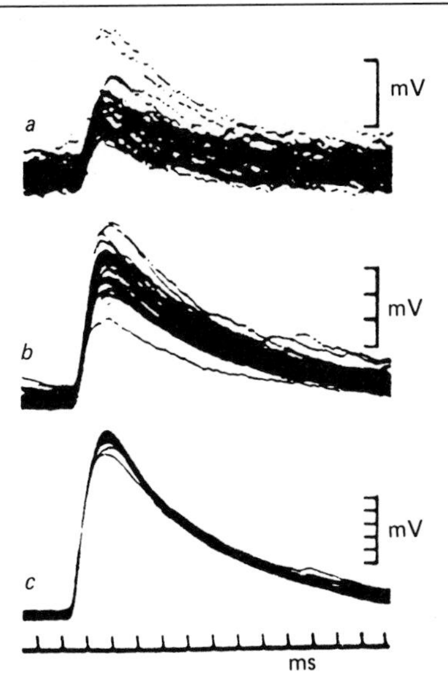

Fig. 8.3. Excitatory postsynaptic potentials (EPSPs) recorded from a cat spinal motoneuron in response to stimuli of increasing intensity (from *a* to *c*) applied to the group Ia afferent fibres from the muscle. From Coombs, Eccles and Fatt (1955*a*).

The postsynaptic responses of motoneurons can be observed by means of a microelectrode inserted into the soma. Stimulation of the group Ia fibres which synapse with a particular motoneuron produces brief depolarizations, as is shown in Fig. 8.3. These responses are called *excitatory postsynaptic potentials*, or EPSPs. Their form is similar to that of the end-plate potential in a curarized skeletal muscle fibre: there is a fairly rapid rising phase followed by a slower return to the resting potential.

Each EPSP is the result of action potentials in a number of presynaptic fibres. With low intensity stimulation applied to the nerve from the muscle, only a few of the group Ia fibres are excited and the EPSP is correspondingly small. As we increase the stimulus intensity, more and more of the group Ia fibres are excited and the EPSP correspondingly grows in size. Thus the responses produced by activity at different synapses on the same motoneuron can add together. This phenomenon is known as *spatial summation*. If a second EPSP is elicited before the first one has died away, the net depolarization will be enhanced as the second EPSP adds to the first. This is known as *temporal summation*.

A large EPSP will be sufficient to cross the threshold for production of an action potential. This then propagates along the axon out to the periphery, where it ultimately produces contraction of the muscle fibres innervated by the axon.

The membrane potential of a motoneuron can be altered by inserting a special double-barrelled microelectrode into it and passing current down one barrel while the other is used to record the membrane potential. When the membrane potential is progressively depolarized, the EPSP decreases in size and eventually becomes reversed in sign. The reversal potential is about 0 mV. This suggests that the EPSP is produced by a change in ionic conductance which is independent of membrane potential, just as is the end-plate potential in muscle. The ions involved are probably sodium and potassium, just as they are in the end-plate potential.

These similarities between the EPSP and the end-plate potential, together with the existence of synaptic vesicles in the presynaptic terminals, suggest that the EPSP is produced by a transmitter substance released from the group Ia terminals. We can assume that there are specific receptors for the transmitter substance, and that these receptors are linked to ionic channels which will allow sodium and potassium ions to flow through them when they are opened. The transmitter substance here is glutamate, not acetylcholine.

INHIBITION IN MOTONEURONS

If the contraction of a particular limb muscle is to be effective in producing movement, it is essential that the muscles which oppose this action (the antagonists) should be relaxed. In the monosynaptic stretch reflex this is brought about by *inhibition* of the motoneurons of the antagonistic muscles. Fig. 8.4 shows the arrangement of the neurons involved. We have seen that group Ia fibres from the stretch receptors in a particular muscle synapse with motoneurons innervating that muscle. They also synapse with small interneurons which themselves innervate the motoneurons of antagonistic

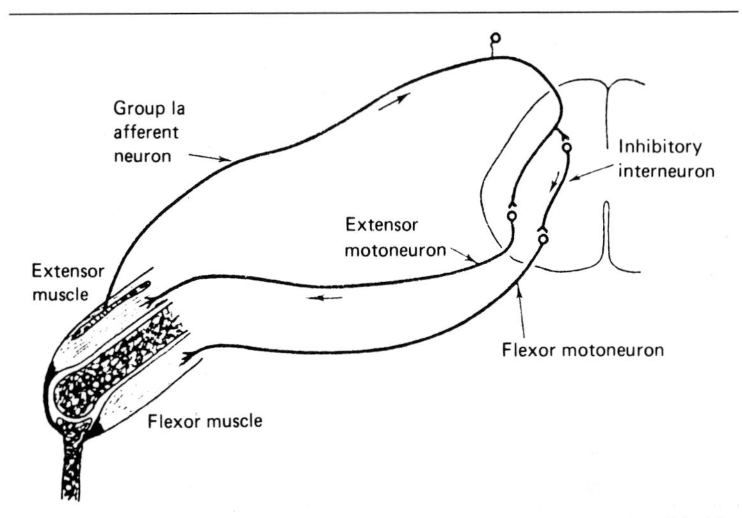

Fig. 8.4. The direct inhibitory pathway. The diagram is much simplified in that there are many afferent, inhibitory and efferent neurons at each stage; each inhibitory interneuron is innervated by several afferents, and itself innervates several motoneurons.

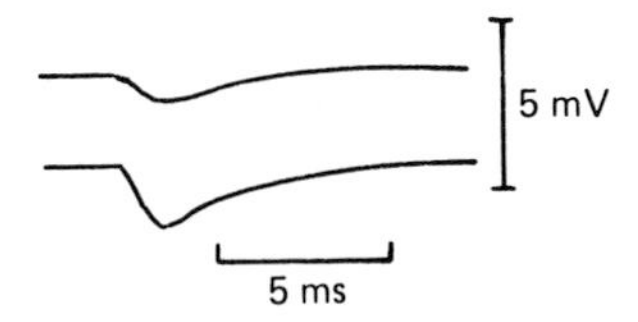

Fig. 8.5. Inhibitory postsynaptic potentials (IPSPs) in a cat spinal motoneuron, produced by stimulating the group Ia fibres from the antagonistic muscle. The stimulus intensity was higher for the lower trace than for the upper, so that more group Ia fibres were excited. From Coombs, Eccles and Fatt (1955*b*).

muscles. It is these interneurons which exert the inhibitory action on the motoneurons.

This inhibitory action can be examined by inserting a microelectrode into a motoneuron and stimulating the group Ia fibres from an antagonistic muscle. Fig. 8.5 shows the results of such an experiment. The responses consist of small hyperpolarizing potentials known as *inhibitory postsynaptic potentials*, or IPSPs.

The form of the IPSP is very similar to that of the EPSP, apart from the fact that it is normally hyperpolarizing. Displacement of the motoneuron membrane potential produces more or less linear changes in the size of the IPSP, with a reversal potential at about −80 mV. This is near to the equilibrium potentials of both chloride and potassium ions. Injection of chloride ions into the soma causes an immediate reduction in the reversal potential, so that the IPSP becomes a depolarizing response at the normal membrane potential. This suggests very strongly that an increase in the chloride ion conductance of the postsynaptic membrane is involved in the production of the IPSP. The situation with potassium ions is less clear.

We can conclude that the IPSP is produced in a way very similar to that of the EPSP and the end-plate potential. An action potential arriving at the presynaptic terminal causes release of a transmitter substance (which is glycine in the spinal cord) from the synaptic vesicles. The transmitter then combines with postsynaptic receptors which are linked to ionic channels which allow chloride ions to flow across the postsynaptic membrane, so producing the IPSP.

IPSPs show spatial and temporal summation just as do EPSPs.

Interaction of IPSPs with EPSPs

The peak depolarization during an EPSP is reduced if there is an overlap in time with an IPSP. If the EPSP was just large enough to elicit an action potential in the absence of the IPSP, then the IPSP may reduce the EPSP so that it no longer crosses the threshold for production of an action potential (Fig. 8.6). When the motoneuron is prevented from producing an action potential in this way it cannot induce muscular contraction and so it is effectively inhibited.

The motoneuron is in a sense a decision-making device. The decision to be made is whether or not to 'fire', that is to say whether or not to send an action potential out along the axon towards the muscle. If the incoming excitatory synaptic action is sufficiently in excess of the incoming inhibitory action, the resulting depolarization will cross the threshold for production of an action potential and the motoneuron will 'fire'. But a reduction in synaptic excitation or an increase in synaptic inhibition will make the membrane potential more negative so that it drops below the threshold and the motoneuron ceases firing. We should remember that the moto-

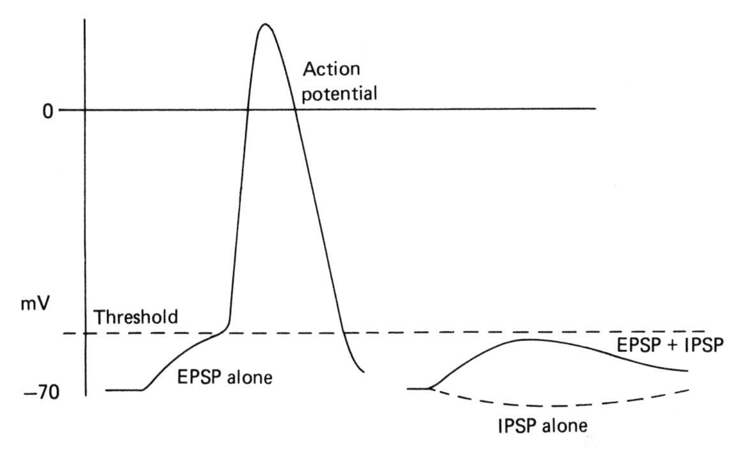

Fig. 8.6. Interaction between excitatory and inhibitory PSPs in the motoneuron. The diagram shows an EPSP which is just large enough to cross the threshold for excitation of an action potential. When an IPSP occurs at the same time, the combined result is insufficient to cause excitation, and so no action potential is propagated out along the axon.

neuron receives excitatory and inhibitory inputs from many sources, so that, for example a 'decision' based on inhibition from group Ia fibres from an antagonistic muscle may be 'overruled' by excitatory inputs for neurons descending from the brain.

Presynaptic inhibition

The inhibitory processes described so far involves the production of hyperpolarizing responses in the postsynaptic cell, and it is thus known as postsynaptic inhibition. Most inhibitory interactions between nerve cells are of this type. In some cases, however, inhibition occurs without there being any postsynaptic response to the inhibitory input alone (Fig. 8.7). This is thought to be caused by synaptic inputs to the presynaptic terminal, which reduce the size of the presynaptic action potential and so reduce the number of transmitter quanta released. Electron microscopy shows the presence of *serial synapses* (Fig. 8.7*c*), in good agreement with this view. The process is known as presynaptic inhibition.

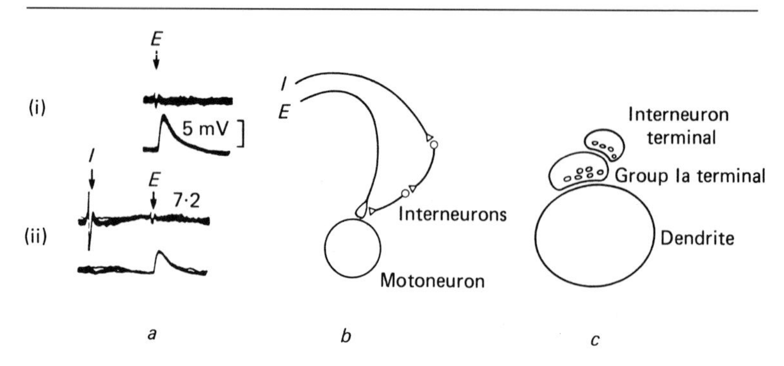

Fig. 8.7. Presynaptic inhibition. *a* (i) shows the EPSP produced in a motoneuron in response to stimulation (at time *E*) of the group Ia fibres innervating it. When a suitable inhibitory nerve is stimulated just beforehand (at *I*), the EPSP is reduced in size although there is no IPSP or other postsynaptic event associated with the inhibitory stimulation, as is shown in *a* (ii). *b* shows the probable nervous pathways and *c* shows the serial synapses which are thought to be involved. *a* From Eccles (1964).

Inhibitory neurotransmitter receptors

Many of the fast synaptic actions in the central nervous system are mediated by the amino acid neurotransmitters glutamate, glycine and gamma-amino butyric acid (GABA). Postsynaptic inhibition is largely mediated by glycine in the spinal cord and by GABA in the brain.

Receptors for GABA and glycine been subjected to recombinant DNA techniques so that we now have some idea of their structure. Each GABA receptor is a multimeric protein consisting of a small number of similar peptide chain subunits. There are at least four different types of these, named α, β, γ and δ; they range in size from 428 to 449 amino acid residues, corresponding to about 50 kDa each. In each chain there are four hydrophobic segments which are likely to be transmembrane α helices. One possible structure for the whole receptor includes two α chains, two β chains and one γ chain, but it seems likely that different combinations of chains are used at different sites in the nervous system.

The glycine receptor also seems to be a multimeric structure. At least one of its component subunits is very similar to the subunits of the GABA receptor, with four transmembrane helices similarly placed in the molecule.

A remarkable and exciting feature of these GABA and glycine receptor subunit sequences is that they show considerable homology with the subunits of the acetylcholine receptor. It looks as though all three receptors are derived from a common evolutionary ancestor. They form a family of neurotransmitter receptors with fast intrinsic ionic channels.

SLOW SYNAPTIC POTENTIALS

Dale used pharmacological criteria to distinguish two types of response to acetylcholine in peripheral tissues. *Nicotinic* responses are mimicked by nicotine and blocked by curare, whereas *muscarinic* responses are mimicked by muscarine and blocked by atropine. Correspondingly, we find there are two distinct types of acetylcholine receptor, nicotinic and muscarinic. Nicotinic receptors occur at the skeletal neuromuscular junction, muscarinic receptors mediate the responses of heart muscle to vagal stimulation. Both types are found in sympathetic ganglia, where they produce different types of responses: let us have a look at them.

The postsynaptic cells in bullfrog sympathetic ganglia show a number of different types of synaptic activity, as is shown in Fig. 8.8. A single stimulus to the preganglionic fibres produces a fast EPSP which may be large enough to produce an action potential in the postganglionic fibres. The response is blocked by curare and can be mimicked by acetylcholine. Thus the mechanism of production of the fast EPSP is similar to that at the neuromuscular junction: it is mediated by nicotinic acetylcholine receptors in which a cation-selective channel opens when acetylcholine is bound to it.

In some cells a slow EPSP with a much longer time course occurs after the fast EPSP. Similar responses are seen after application of acetylcholine. The slow EPSP is unaffected by curare but is blocked by atropine, hence the receptors which mediate it are muscarinic. Conductance measurements show that the slow EPSP is produced by the closure of ionic channels selective for potassium ions.

In other cells the fast nicotinic EPSP is followed by a slow,

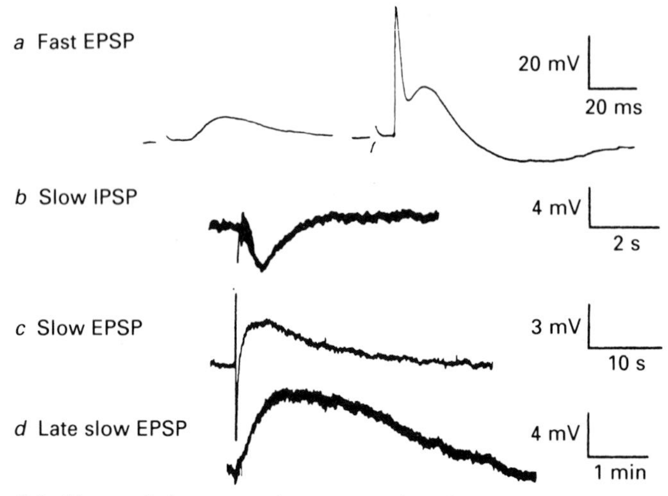

Fig. 8.8. Fast and slow synaptic responses in a frog sympathetic ganglion neurons. The trace on the left in *a* shows the fast EPSP produced by a single preganglionic stimulus; a stronger stimulus (right) excites more preganglionic fibres giving a larger EPSP which is sufficient to produce an action potential. In *b* to *d* the fast EPSP is blocked by a curare-like compound; repetitive stimulation at various sites then produces three different types of slow response. Note the different time scales. From Kuffler (1980).

hyperpolarizing IPSP. This also is muscarinic, and probably involves the opening of potassium channels. Finally, a long period of repetitive stimulation of the preganglionic fibres produces a depolarization which lasts for a few minutes; it is called the late slow EPSP. The neurotransmitter which produces this is a peptide similar in structure to the luteinizing hormone-releasing hormone.

Slow potentials are widely distributed. Their time course and their long latency could be explained if channel opening or closing is mediated by an indirect process involving intermediate steps between binding at the receptor and the response of the channel, rather than the direct link which occurs in fast-acting receptors with intrinsic channels. The intermediate steps involve the activation of G proteins and often the production of intracellular 'second messengers'.

The second messenger concept was first introduced to describe the role of cyclic adenosine monophosphate (cyclic AMP) in hormone action. Combination of a hormone with its receptor activates a G protein (so called because it needs to bind guanosine triphosphate to become active) which in turn activates the enzyme adenylate cyclase. This produces cyclic AMP which then alters the physiological properties of the cell in some way, such as by opening or closing channels. Neurotransmitters may act in a similar fashion, or may utilise a different second messenger such as inositol trisphosphate. In some cases the G protein may act directly on the membrane channel without producing a second messenger. Fig. 8.9 summarizes the various ways in which neurotransmitters may affect channels.

G protein linked receptor structures

The best known of the receptors which act via G proteins are the muscarinic acetylcholine receptor and the β-adrenergic receptor. The latter mediates many of the responses to noradrenaline in smooth and heart muscle cells.

Molecular cloning techniques have shown that the muscarinic acetylcholine receptor and the β-adrenergic receptor are strikingly similar in structure, with identical amino acids at 30% of their residues. Their amino acid sequences are also surprisingly similar to that of the visual pigment rhodopsin, with 23% homology in each case. All three molecules have seven membrane-crossing helices, as is shown in Fig. 8.10. Here, then, we have a separate family of receptors which have a quite different structure from those which produce fast responses involving the opening of an intrinsic channel.

The chain lengths in this family are 460 amino acids for the muscarinic receptor, 418 for the β-adrenergic receptor and 348 for rhodopsin. Molecules of this size are probably not large enough to accommodate an intrinsic channel (compare the values of 1820 amino acids for the sodium channel and a total of 2333 for the nicotinic acetylcholine receptor) and there is no indication that multimeric associations occur. Instead, they are closely linked to the G proteins which act as shuttles between the receptors and the channels which they ultimately activate.

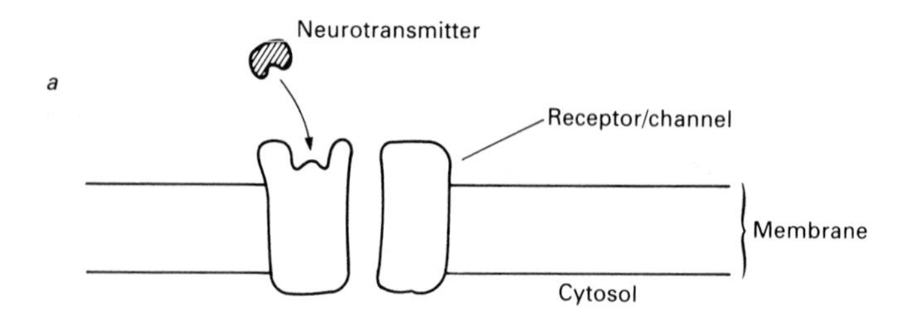

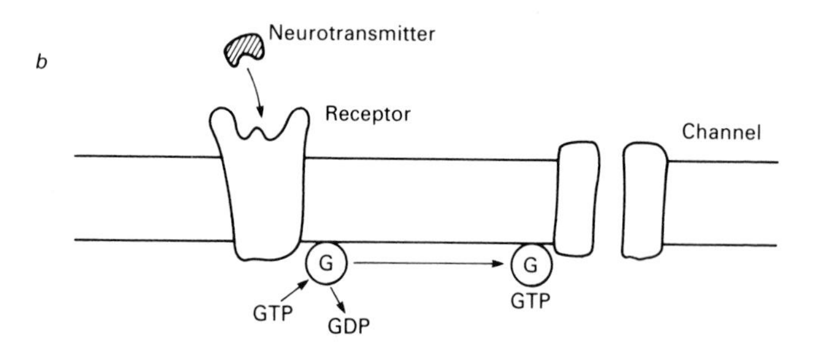

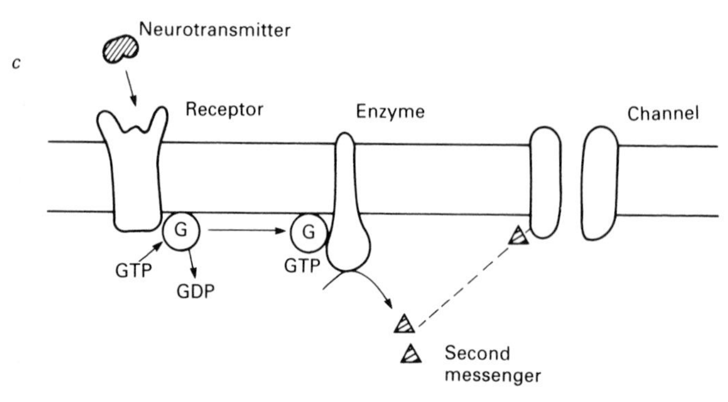

Fig. 8.9. Direct and indirect action of neurotransmitters on ionic channels. *a* shows the direct action which occurs when the ionic channel is an integral part of the receptor, as in the nicotinic acetylcholine receptor. In *b* and *c* the receptor molecule acts indirectly via activation of a G protein. In *b* the G protein acts directly on the channel to open or close it. In *c* the G protein activates an enzyme which generates a second messenger such as cyclic AMP which itself then alters the state of the channel. From Aidley (1990).

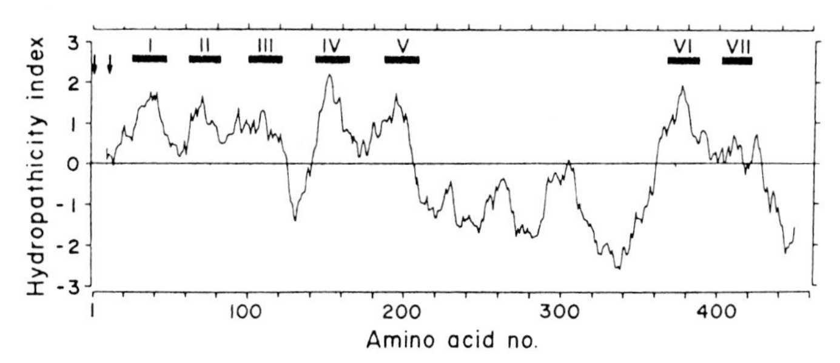

Fig. 8.10. Hydropathicity profile of the muscarinic acetylcholine receptor, calculated from its amino acid sequence. The graph shows the average hydropathicity index (see Table 5.1) of the 19 amino acids from $i-9$ to $i+9$, plotted against i, where i represents the amino acid number. Black bars show the positions of the putative transmembrane segments I to VII. From Kubo *et al.* (1986).

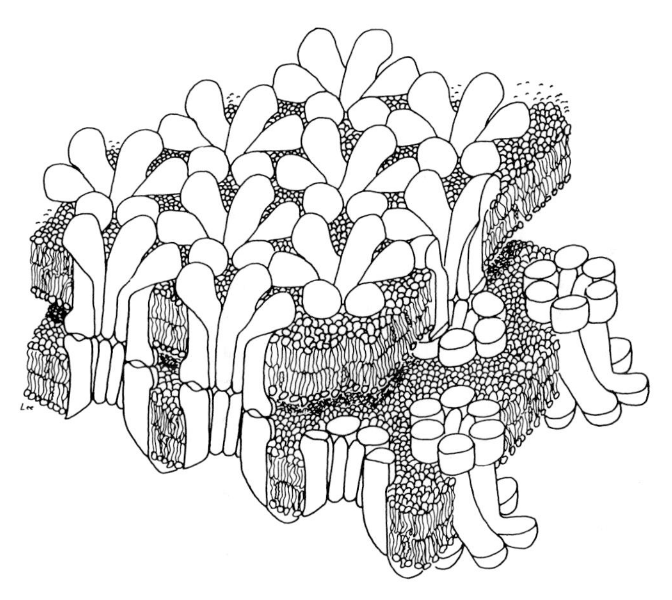

Fig. 8.11. The structure of gap junctions, as deduced from X-ray diffraction studies on material isolated from mouse liver cells. From Makowski *et al.* (1984).

ELECTROTONIC SYNAPSES

An electrotonic synapse is one where the presynaptic cell excites the postsynaptic cell directly by means of electric current flow. Synapses of this type were first discovered in the multicellular giant fibres involved in the escape responses of crayfish and earthworms, where they serve to carry the action potential from one cell to another. They also occur between some cells in the central nervous systems of mammals and other animals, where they probably serve to promote synchrony of action in adjacent cells.

Electron microscopy of electrotonic synapses shows regions where the intercellular space between the two cells is much narrower than usual, being about 2 nm instead of 20 nm. These regions are known as 'gap junctions'. They contain channels which provide direct connections between the pre- and postsynaptic cells, so that current can flow readily from one cell to the other. Each gap junction channel is composed of a pair of hexamers ('connexons'), one in each of the two apposed membranes, as is shown in Fig. 8.11.

CHAPTER NINE

Skeletal muscles

Skeletal muscles are the engines of the body. They account for over a quarter of its weight and the major part of its energy expenditure. They are attached to the bones of the skeleton and so serve to produce movements or exert forces. Hence they are used in such activities as locomotion, maintenance of posture, breathing, eating, directing the gaze and producing gestures and facial expressions.

Skeletal muscles are activated by motoneurons as we have seen in previous chapters. Their cells are elongate and multinuclear and the contractile material within them shows cross-striations, hence skeletal muscle is a form of *striated muscle*. In contrast, cardiac and smooth muscles have cells with single nuclei, and smooth muscles are not striated; we shall examine their properties in a later chapter.

ANATOMY

Skeletal muscle fibres are multinucleate cells (Fig. 9.1) formed by the fusion of numbers of elongated uninucleate cells called myoblasts. Mature fibres may be as long as the muscle of which they form part, and 10 to 100μm in diameter. The nuclei are arranged around the edge of the fibre. Most of the interior of the fibre consists of the protein filaments which constitute the contractile apparatus, grouped together in bundles called *myofibrils*. The myofibrils are surrounded by cytoplasm (or *sarcoplasm*), which also contains mitochondria, the internal membrane systems of the sarcoplasmic reticulum and the T system, and a fuel store in the form of glycogen granules and sometimes a few fat droplets. We shall examine the structure of the contractile apparatus and the internal membrane systems in more detail in the following chapter.

The muscle fibre is bounded by its cell membrane, sometimes called the sarcolemma, to which a thin layer of connective tissue (the

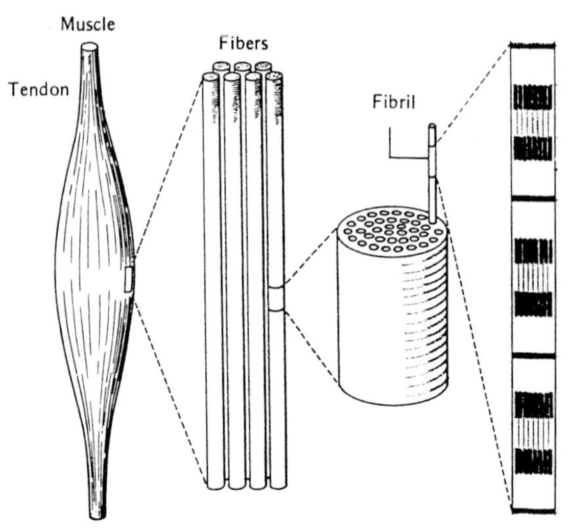

Fig. 9.1. Diagram to show the arrangement of fibres in a vertebrate striated muscle. The cross-striations on the myofibrils can be seen with light microscopy; their ultrastructural basis is shown in Fig. 10.6. After Schmidt-Neilsen (1990).

endomysium) is attached. Bundles of muscle fibres are surrounded by a further sheet of connective tissue (the perimysium) and the whole muscle is contained within an outer sheet of tough connective tissue, the epimysium. These connective tissue sheets are continuous with the insertions and tendons which serve to attach the muscles to the skeleton.

Muscles have an excellent blood supply, with blood capillaries forming a network between the individual fibres. Sensory and motor nerve fibres enter the muscle in one or two nerve branches. The sensory nerve endings include those on the muscle spindles (sensitive to length), in the Golgi tendon organs (sensitive to tension), and a variety of free nerve endings in the muscle tissue, some of which are involved in sensations of pain.

In mammals the gamma motoneurons provide a separate motor nerve supply for the muscle fibres of the muscle spindles, while the bulk of the muscle fibres are supplied by the alpha motoneurons. Each alpha motoneuron innervates a number of muscle fibres, from less than ten in the extraocular muscles (those which move the eyeball in its socket) to over a thousand in a large limb muscle. The complex of one motoneuron plus the muscle fibres which it innervates is called a *motor unit*. Since they are all activated by the same nerve cell, all the muscle fibres in a single motor unit contract at the same time. Muscle fibres belonging to different motor units may well contract at different times, however.

Most mammalian muscle fibres are contacted by a single nerve terminal, although sometimes there may be two terminals originating from the same nerve axon. Muscle fibres of this type are known as *twitch fibres*, since they respond to nervous stimulation with a rapid twitch. In the frog and other lower vertebrates another type of muscle fibre is commonly found, in which there are a large number of nerve terminals on each muscle fibre. These are known as *tonic fibres*, since their contractions are slow and maintained. There are some tonic fibres in the extraocular muscles of mammals, and also in the muscles of the larynx and the middle ear.

MECHANICAL PROPERTIES

The mechanical properties of muscles are best investigated with isolated muscle or nerve–muscle preparations such as the gastrocnemius or sartorius nerve–muscle preparations in the frog. Experiments on large mammalian muscles require an intact blood supply, in which case the experiments must be performed on an anaesthetized animal, with the nerve supply to the muscle cut and its tendon dissected free and attached to some recording device. The muscle is excited by applying a brief pulse of electric current to its nerve or directly to the muscle itself.

Isometric and isotonic contractions

When muscles contract they exert a force on whatever they are attached to (this force is equal to the *tension* in the muscle) and they shorten if they are permitted to do so. Hence we can measure two different variables during the contraction of a muscle: its length and its tension. Most often one of these two is maintained constant during the contraction. In *isometric* contractions the muscle is not allowed to shorten (its length is held constant) and the tension it

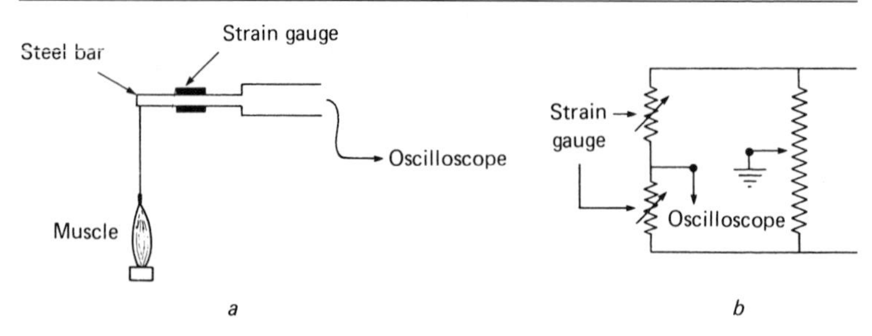

Fig. 9.2. An isometric lever system for measuring the force exerted by a muscle without allowing it to shorten. Semiconductor strain gauges are bonded to a steel bar (*a*), and form two arms of a resistance bridge connected to a battery (*b*).

produces is measured. In an *isotonic* contraction the load on the muscle (which is equal to the tension in the muscle) is maintained constant and its shortening is measured.

An isometric recording device has to be stiff, so that it does not in fact allow the muscle to shorten appreciably while the force is being measured. A simple method is to use a lever which is attached to a stiff spring and writes on a smoked drum. A more sophisticated device consists of a small steel bar to which semiconductor strain gauges are bonded. The electrical resistance of the strain gauges then varies with muscle tension and so can be used to give an electrical measure of the tension, and this can then be displayed on an oscilloscope or a chart recorder (Fig. 9.2). The force exerted by the muscle is usually measured in newtons or grams weight.

Isotonic recording devices usually consist of a moveable lever whose motion can be recorded either directly on a smoked drum or indirectly via an electrical signal. The lever can be loaded to different extents, perhaps by hanging weights on it. Fig. 9.3 shows a typical arrangement.

Isometric twitch and tetanus

When a single high intensity stimulus is applied to a muscle arranged for isometric recording, there is a rapid increase in tension which then decays away (Fig. 9.4). This is known as a *twitch*. The duration of the twitch varies from muscle to muscle, and decreases

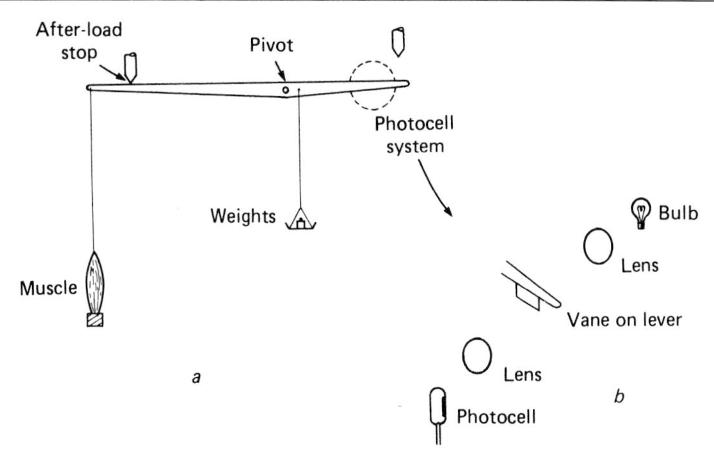

Fig. 9.3. An isotonic lever system (*a*); and (*b*) the photocell system used to record the position of the lever.

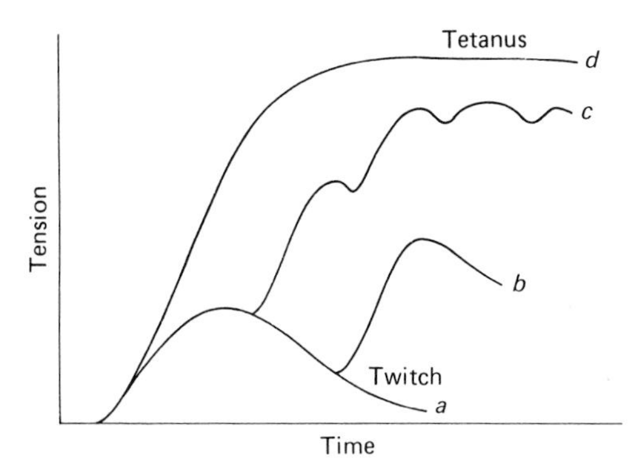

Fig. 9.4. Isometric contractions. *a*, response to a single stimulus, producing a twitch; *b*, response to two stimuli, showing mechanical summation; *c*, response to a train of stimuli, showing an 'unfused tetanus'; *d*, response to a train of stimuli at a higher repetition rate, showing a maximal fused tetanus.

with increasing temperature. For a frog sartorius at 0 °C, a typical value for the time between the beginning of the contraction and its peak value is about 200 ms, tension falling to zero again within 800 ms.

If a second stimulus is applied before the tension in the first twitch has fallen to zero, the peak tension in the second twitch is higher than that in the first. This effect is known as *mechanical summation*. Repetitive stimulation at a low frequency thus results in a 'bumpy' tension record. As the frequency of stimulation is increased, a point is reached at which the bumpiness is lost and the tension rises smoothly to reach a steady level. The muscle is then in *tetanus*, and the minimum frequency at which this occurs is known as the *fusion frequency*.

The ratio of the peak tension in an isometric twitch to the maximum tension in a tetanus is called the twitch/tetanus ratio. It may be about 0.2 for mammalian muscles at 37 °C, and rather higher for frog muscles at room temperature or below.

A low intensity stimulus applied to the nerve may produce no contraction of the muscle; this is because the current flow is too small to excite any of the nerve fibres. As we increase the intensity of the stimulus, more and more nerve fibres are excited, so that more and more motor units are activated and hence the total tension gets greater and greater. Eventually the stimulus is of high enough intensity to excite all the nerve fibres and so all the muscle fibres are excited; further increase in stimulus intensity does not increase the tension reached by the muscle. Thus the muscle reaches its maximum tension when all its individual muscle fibres are active simultaneously.

This mechanism provides the way in which gradation of muscular force is achieved in the body. Gentle movements involve the simultaneous activity of a small number of motor units, whereas in vigorous movements many more motor units are active.

Mammalian muscles are of two distinct types, called *fast-twitch* and *slow-twitch*. The fast-twitch muscles contract and relax more

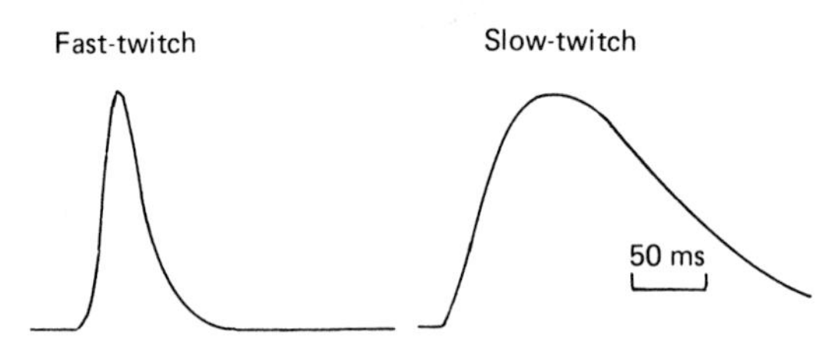

Fig. 9.5. Isometric twitches of two types of cat muscle, showing the much longer time course of the slow-twitch muscle. From Buller (1975).

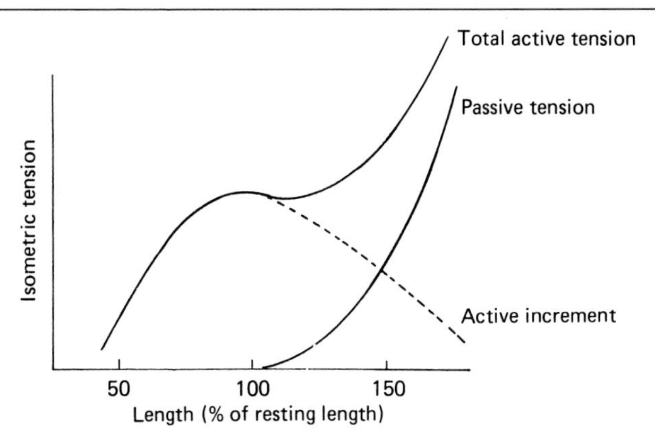

Fig. 9.6. The length–tension relation of a skeletal muscle.

rapidly than the slow-twitch muscles, as is shown in Fig. 9.5. Fast-twitch muscles are used in making fairly rapid movements, whereas the slow-twitch muscles are utilised more for the long-lasting contractions involved in the maintenance of posture. The gastrocnemius, for example, is a fast-twitch muscle used to extend the ankle joint in walking and running, whereas the soleus is a slow-twitch muscle which acts similarly on the same joint while its owner is standing still.

When a resting muscle is stretched it becomes increasingly resistant to further extension, largely because of the connective tissue which it contains. Hence it is possible to determine a passive *length–tension curve*, as is shown in Fig. 9.6. The full isometric tetanus tension of the stimulated muscle is also dependent on length, as is shown in the 'total active tension' curve in Fig. 9.6. The difference between the two curves can be called the 'active increment' curve; notice that this reaches a maximum at a length near to the maximum length in the body, falling away at longer or shorter lengths. We shall see in the next chapter that this has important implications for the nature of the contractile mechanism.

Isotonic contractions

Fig. 9.3 shows a common arrangement for recording isotonic contractions. The after-load stop serves to support the load when

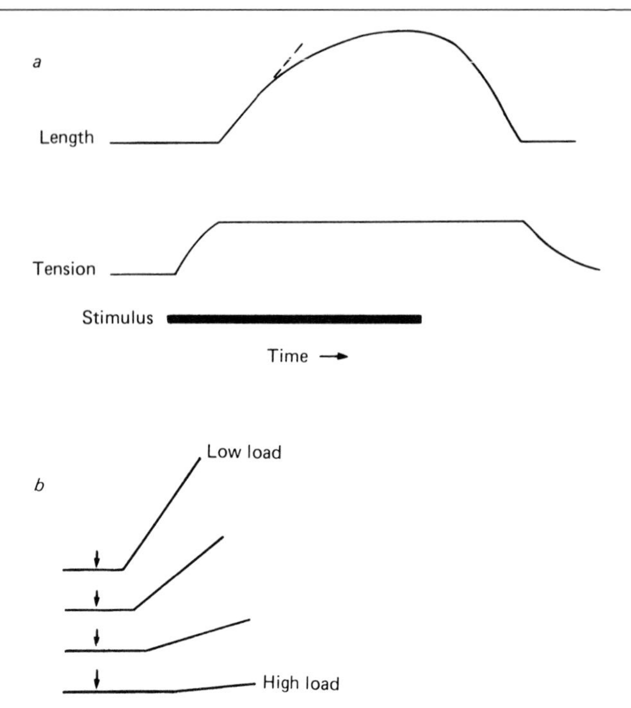

Fig. 9.7. After-loaded isotonic tetanic contractions. *a* shows the length and tension changes during a single contraction, with shortening as an upward deflection of the length trace. *b* shows the initial length changes in contractions against different loads.

the muscle is relaxed and to determine the initial length of the muscle. If it were not there the muscle would take up longer initial lengths with heavier loads, which would make it more difficult to interpret the results of experiments with different loads.

Fig. 9.7*a* shows what happens when the muscle has to lift a moderate load while being stimulated tetanically. The tension in the muscle starts to rise soon after the first stimulus, but it takes some time to reach a value sufficient to lift the load, so that there is no shortening at first and the muscle is contracting isometrically. Eventually the tension becomes equal to the load and so the muscle begins to shorten; the tension remains constant during this time and the muscle contracts isotonically. It is noticeable that initially the velocity of shortening during the isotonic phase is constant, provided that the muscle was initially at a length near to its maximum length in the body. As the muscle shortens further, however, its velocity of shortening falls until eventually it can shorten no further and shortening ceases. When the period of stimulation ends, the muscle is extended by the load as it relaxes until the lever meets the after-load stop, after which relaxation becomes isometric and the tension in the muscle falls to its resting level.

If we repeat this procedure with different loads (as in Fig. 9.7*b*), we find that the contractions are affected in three ways.

(1) The delay between the stimulus and the onset of shortening is longer with heavier loads. This is because the muscle takes longer to reach the tension required to lift the load.
(2) The total amount of shortening decreases with increasing load. This is because the isometric tension falls at shorter lengths (Fig. 9.6) and so the more heavily loaded muscle can only shorten by a smaller amount before its isometric tension becomes equal to the load. Fig. 9.8 illustrates this point.
(3) During the constant-velocity section of the isotonic contraction, the velocity of shortening decreases with increasing load. It becomes zero when the load equals the maximum tension which can be reached during an isometric contraction of the muscle at that length.

Notice that the first two of these observations are essentially predictable from what we already know about isometric contractions. We can quantify the third observation by plotting the initial velocity of shortening against the load to give a *force–velocity curve*, as in Fig. 9.9. The curve is more or less hyperbolic in shape, and is believed by many physiologists to be of fundamental importance in the understanding of muscle functioning.

In 1938 A. V. Hill produced an equation to describe the form of the force-velocity curve, as follows:

$$(P+a)(V+b) = b(P_0-a)$$

or

$$(P+a)(V+b) = \text{constant}$$

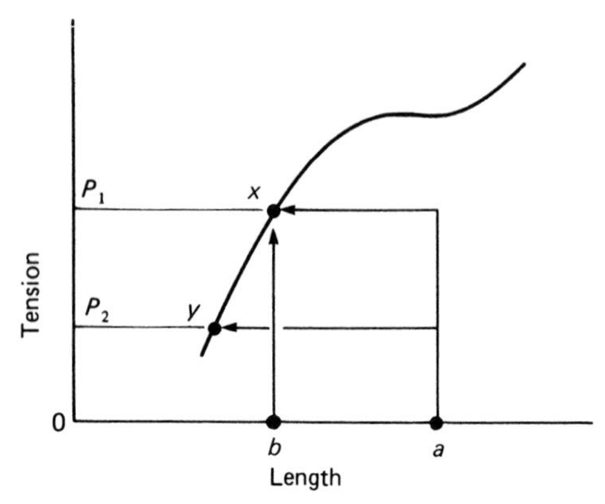

Fig. 9.8. Diagram showing why it is that a lightly loaded muscle can shorten further than a heavily loaded one. Starting from point *a* on the length axis, the muscle contracts isometrically until its tension is equal to the load it has to lift, and then it shortens until it meets the isometric length–tension curve. With a heavy load (P_1) this occurs at *x*, with a lighter load (P_2) it occurs at *y*. Notice that point *x* can also be reached by an isometric contraction from point *b*. (When starting from a much extended length, a muscle may in practice stop short of point *x* when lifting load P_1; this is probably caused by inequalities in the muscle.)

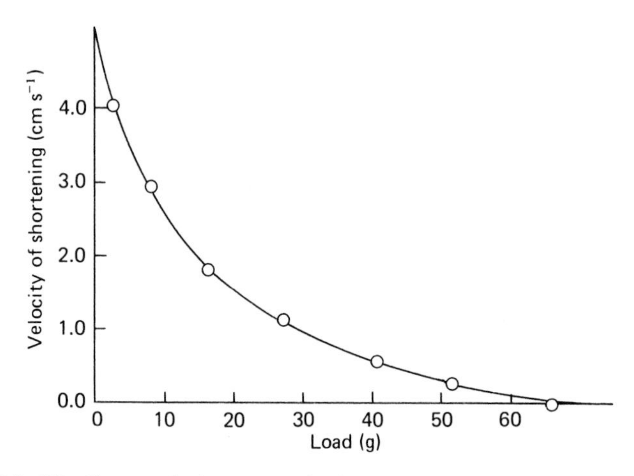

Fig. 9.9. The force–velocity curve of a frog sartorius muscle at 0° C. From Hill (1938).

where V is the velocity of shortening, P is the force exerted by the muscle, P_0 is the isometric tension, and a and b are constants.

ENERGETICS OF CONTRACTION

Energetics is the study of energy conversions. In a contracting muscle chemical energy is converted into mechanical energy (work), with heat energy being produced as a by-product. The law of the conservation of energy suggests that the chemical energy released in a contraction will be equal to the work done plus the heat given out in that contraction. That is to say,

$$\text{chemical energy release} = \text{heat} + \text{work}$$

Of the three terms in this equation, the work is the easiest to measure, then the heat, and the chemical change is the most difficult. Let us examine them in this order.

Work and power

Mechanical energy is measured as *work* and the rate of performing work is called *power*. In an isotonic contraction, therefore, the work done is equal to the force exerted by the muscle multiplied by distance shortened, and the power output at any instant is equal to the force multiplied by the velocity of shortening.

There is no work done by the muscle as a whole during an isometric contraction since there is no shortening, and there is none done during an unloaded isotonic contraction since the force exerted is zero. Power will also be zero under both these conditions. Fig. 9.10. shows how the work done in isotonic twitches (measured between the onset of contraction and its peak) varies with the load. During the relaxation process in an isotonic twitch, the load does work on the muscle and so the overall work output during the whole of the isotonic twitch is zero.

The power output during the initial shortening phase of tetanic isotonic contractions is readily calculated from the force–velocity curve, since power is force times velocity. As shown in Fig. 9.11, maximum power is achieved when the load is about 0.3 times the isometric tension, when the muscle will shorten at about 0.3 times

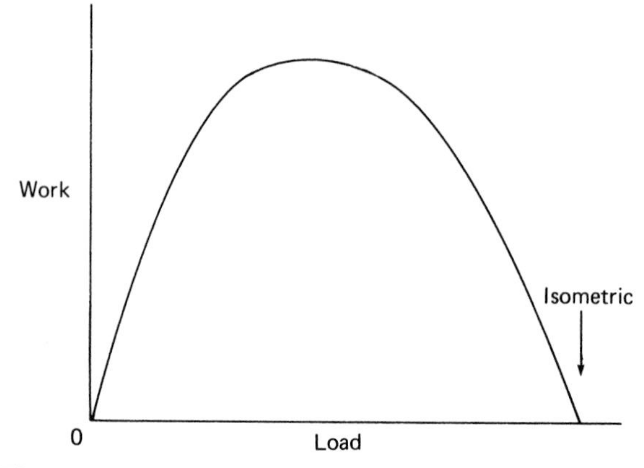

Fig. 9.10. How work varies with load in isotonic twitches.

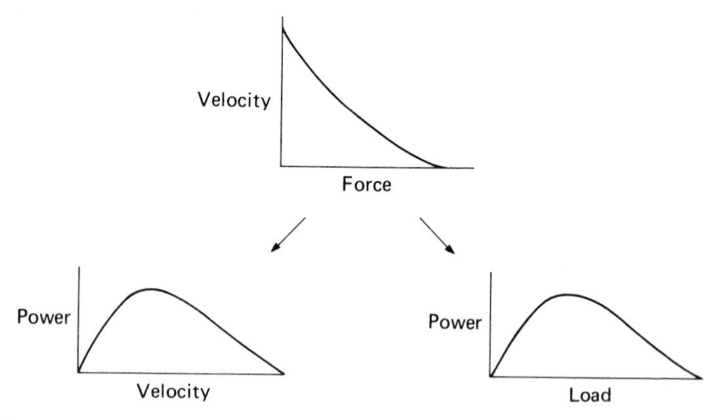

Fig. 9.11. The power output during tetanic isotonic contractions plotted against different loads and velocities.

its maximum (unloaded) velocity. This has implications for the selection of gears in cycle races: whatever the speed of the cycle, maximum power is obtained from the leg muscles when they are contracting at about 0.3 times their maximum unloaded velocity, which probably corresponds to about two revolutions of the pedals per second.

Heat production

Our everyday experience demonstrates that muscular activity is accompanied by heat production. It is necessary to take account of this when studying the energy released by muscle. Much of our knowledge is based on that acquired by A. V. Hill and his colleagues from many years' work on isolated frog muscles. In most experiments the muscle is laid over a thermopile – an array of thermocouples arranged in series – so that very small changes in temperature can be measured.

During an isometric tetanus, heat is released at a very high rate for the first 50 ms or so (this is usually called the *activation heat*), falling rapidly to a lower more steady level which is usually called the *maintenance heat* (Fig. 9.12). If the muscle is allowed to shorten, an extra amount of heat is released during the shortening process. This *heat of shortening* is roughly proportional to the distance shortened. Further heat appears during relaxation, especially if the load does work on the muscle. Finally, after relaxation, there is a prolonged *recovery heat* as the muscle metabolism restores the chemical situation in the muscle to what it was before the contraction took place.

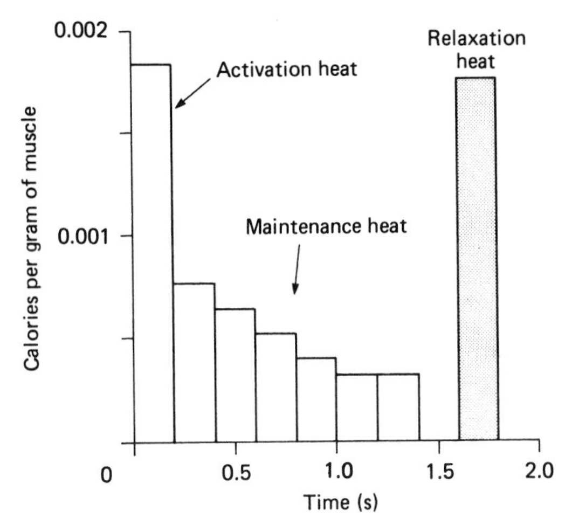

Fig. 9.12. The rate of heat production of frog sartorius muscle during an isometric tetanus. The muscle was stimulated for a period of 1.2 s. From Hill and Hartree (1920).

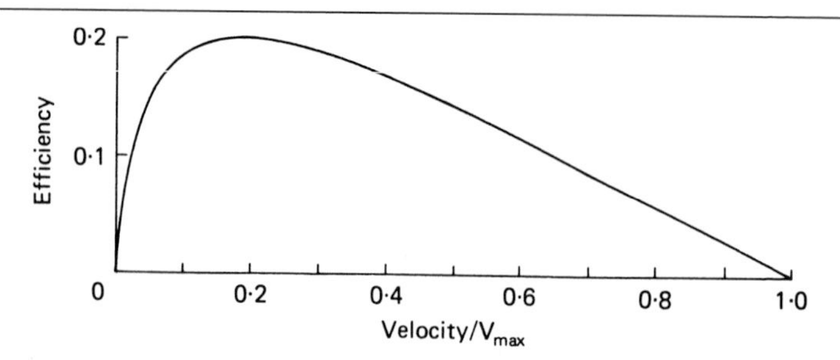

Fig. 9.13. How efficiency varies with velocity during isotonic tetanic contractions. From Hill (1950*b*).

Efficiency

The efficiency of a muscle is a measure of the degree to which the energy expended is converted into work, i.e.

$$\begin{aligned} \text{efficiency} &= \text{work}/(\text{total energy release}) \\ &= \text{work}/(\text{heat}+\text{work}) \end{aligned}$$

If we consider only the energy changes during and immediately after the contraction, the efficiency works out at about 0.4 for frog muscle and 0.8 for tortoise muscle. But if we include the recovery heat in the calculation, the lower values of 0.2 and 0.35 are obtained. These are maximum values, obtained by allowing the muscle to shorten at about one fifth of its maximum velocity. A. V. Hill's calculation of how efficiency varies with velocity of shortening is shown in Fig. 9.13. Notice that maximum efficiency occurs at a lower velocity than that at which maximum power output occurs. However both curves have fairly broad peaks so the difference between them may not be very important.

The energy source

So far we have concentrated on the output side of the energy balance equation. It is now time to consider the question, what chemical changes supply the energy for muscular contraction?

Energy for all bodily activities is ultimately derived from the food. Food energy is transported to the muscle as glucose or fatty acids and may be stored there as glycogen (a polymer of glucose).

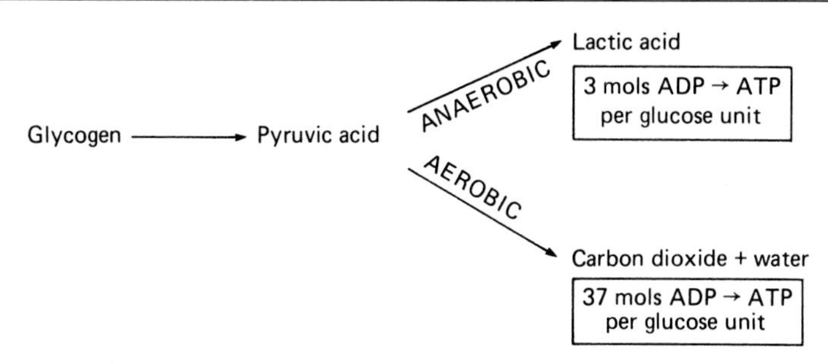

Fig. 9.14. An outline of the breakdown of glycogen with the release of energy (in the form of ATP) in respiration.

Respiration of these substances within the muscle cells results in the production of adenosine triphosphate (ATP) from adenosine diphosphate (ADP), as is indicated in Fig. 9.14. ATP appears to be the immediate source of energy for a large number of cellular activities.

The 'high energy phosphate' can be transferred from ATP to creatine (Cr), forming creatine phosphate (CrP):

$$\mathrm{ATP + Cr \longrightarrow ADP + CrP}$$

This reaction is catalysed by the enzyme creatine phosphotransferase; it is readily reversible, so that the creatine phosphate can form a short-term 'bank' of high energy phosphate.

This scheme is now familiar to all who study elementary biochemistry, but it is worth examining some of the evidence that it applies to muscle. In 1950 A. V. Hill issued a famous 'challenge to biochemists' in which he said that if ATP really was the immediate source of chemical energy, then it should be possible to demonstrate that ATP was broken down during a contraction in living muscle.

The general method used in the experiments that followed was to use metabolic inhibitors to prevent the resynthesis of high energy phosphate, and then to compare its concentration in two muscles of which only one had been stimulated. The muscles had to be very rapidly frozen after the contraction (by plunging them into liquid propane, for example) so as to prevent any further biochemical change.

R. E. Davies and his colleagues used the substance 1-fluoro-2,4-

dinitrobenzene (FDNB) to block the action of creatine phosphotransferase in frog muscles, so that ADP could not be rephosphorylated to ATP by the breakdown of creatine phosphate. In one set of experiments they found that the stimulated muscles lost on average 0.22 μmoles of ATP per gram of muscle in an isotonic twitch. Now the heat of hydrolysis of the terminal phosphate bond of ATP is about 34 kJ/mole, so the breakdown of 0.22 μmoles should release about 7.5×10^{-3} J. The work done by the muscle amounted to 1.7×10^{-3} J per gram of muscle. This means that the ATP breakdown is more than sufficient to account for the work done in the twitch; the excess energy appears as heat.

Another way of measuring the 'fuel consumption' of the muscle is to measure the differences in creatine phosphate content of stimulated and unstimulated muscles. Here FDNB is not used because one wishes the transfer of phosphate from creatine phosphate to ATP to occur rapidly, as in the normal muscle. The resynthesis of creatine phosphate is prevented by poisoning the muscle with iodoacetate, which blocks one of the enzymic reactions in the breakdown of glycogen, in an atmosphere of nitrogen. Under these conditions there is less creatine phosphate in the stimulated muscle than in the unstimulated one.

It has proved quite difficult to draw up a precise balance sheet for the energy changes in muscle. D. R. Wilkie made some careful measurements on the energy output (heat + work) and creatine phosphate breakdown in frog muscles during a variety of different types of contraction. His results (Fig. 9.15) showed that the energy output is linearly proportional to the breakdown of creatine phosphate, with 46.4 kJ of energy being produced for each mole of creatine phosphate broken down.

However, calorific measurements suggest that the hydrolysis of creatine phosphate should yield only about 32 kJ per mole. The gap between expectation and observation became known as the 'unexplained energy'. More recent results (discussed by Homsher, 1987) suggest that at least part of this is connected with calcium-binding reactions associated with the activation processes in the muscle, but there are still some features of muscle energy balance which are not understood.

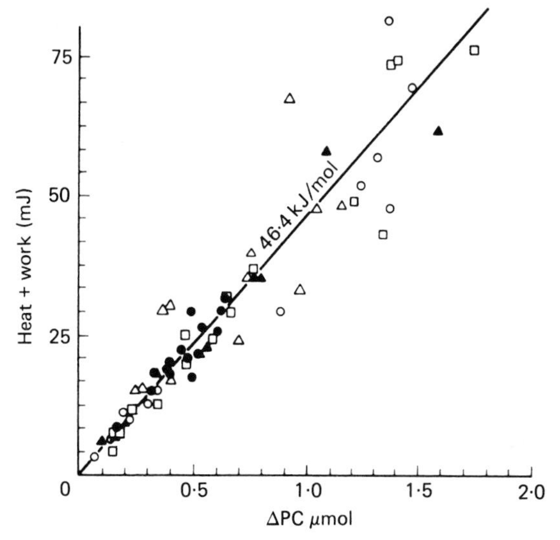

Fig. 9.15. The relation between energy production (heat plus work) and creatine phosphate breakdown in frog sartorius muscles poisoned with iodoacetate and nitrogen. Each point represents a determination on one muscle after the end of a series of contractions, with different symbols for different types of contraction. From Wilkie (1968).

MUSCULAR EXERCISE

The speed at which a man can run is determined in part by the speed with which he can convert his stores of chemical energy into mechanical energy. These stores may be within the muscle or outside it, and they may or may not require oxygen from outside the cell in order to be utilized. The three main energy stores are as follows.

(1) ATP and creatine phosphate in the muscle. This is the short term energy store, amounting to about 16 kJ or so in the human body, perhaps enough for a minute of brisk walking.

(2) Glycogen in the muscle and the liver. This provides a medium term store of very variable size: a value of 4000 kJ would not be out of the ordinary, providing enough energy for some hours of moderate exercise.

(3) Fat in the adipose tissue. This provides a long term store; 300000 kJ might be a typical value.

The high energy phosphate store can be immediately utilised by the contractile apparatus of the muscle. The energy in the fat store and most of that in the glycogen store can only be utilised by aerobic respiration, for which oxygen has to be transported to the cell. The rate at which these two energy stores can be tapped is therefore limited by the rate at which oxygen can be supplied to the cell. It is for this reason that the maximum running speeds for short distances cannot be maintained over long distances (Fig. 9.16).

The muscles' increased need for oxygen during steady exercise is served by the well-known physiological changes which occur in the body during exercise. The rate and depth of breathing increases, the heart rate and stroke volume increase, and the blood supply to the muscles is increased. The concentration of free fatty acids in the blood rises, as a result of the hydrolysis of some of the fat stored in the adipose tissue. There is also some mobilisation of the glycogen reserves of the liver.

Some of the glycogen in the muscle can contribute to the short term energy store since it can be utilised without oxygen supplied

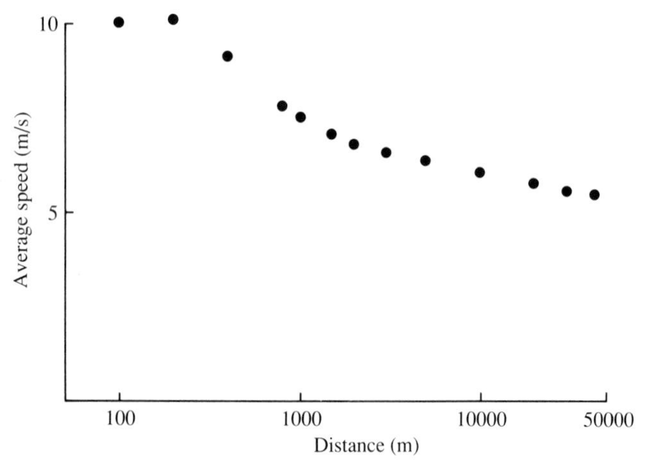

Fig. 9.16. How the average speed varies with distance run. The points are determined from world records for men as they were in October 1987, with the distances plotted on a logarithmic scale.

rom outside. Anaerobic respiration, producing lactic acid, supplies moles of ATP per glucose unit used. This is a much less effective rocess than aerobic respiration, which supplies 37 moles of ATP for each glucose unit. There is also some oxygen bound to myoglobin in the muscle; this might amount to 0.4 l or so, enough for a few seconds of maximal exercise.

Muscular fatigue

The force exerted during a maximal voluntary isometric contraction in man begins to decline after a few seconds. A similar contraction in which the muscle exerts a tension of 50% of the maximum can be maintained for about a minute, and one of 15% for more than ten minutes. This inability to maintain the tension at a particular level is called *fatigue*. It is usually accompanied by feelings of discomfort and perhaps pain in the muscle.

Experiments by P. A. Merton and his colleagues on the adductor muscle of the thumb suggest that fatigue is largely a feature of the contractile machinery of the muscle cell. The force in maximal voluntary isometric contractions declined to less than half after 1 to 3 minutes, but there was no reduction in the size of the muscle action potential produced by electrical stimulation of the motor nerve. Maximal force could not be restored by massive direct stimulation of the muscle fibres.

It seems likely that the causes of fatigue are connected with changes in the fuel supply for the contractile machinery of the muscle; the precise details are not yet clear. Fatigue may well act in part as a protective mechanism, avoiding the reduction of ATP concentrations to dangerously low levels: rigor mortis occurs in the absence of ATP.

Fatigue is more rapid for isometric than for isotonic contractions. Strong isometric contractions result in increased pressure in the muscle, which reduces and may completely block the blood flow through it. Consequently the muscle runs out of oxygen and hence its energy supply is soon used up. Rhythmic contractions can be maintained for a much longer time since they produce only intermittent interruptions in the blood supply.

The effects of training

Muscles are markedly affected by the amount of use they receive. The structural and biochemical changes in human muscles can be investigated by the technique of needle biopsy, whereby a small piece of muscle is removed for analysis.

Muscles can be increased in size and strength by exercise involving the development of high muscle tensions, such as in isometric exercises or weightlifting. These procedures result in enlargement of the individual fibres in the muscle, including an increase in the quantity of contractile protein in them.

Training of the muscles by endurance exercises, as in distance running for example, results in an appreciable increase in the blood supply to the muscle via proliferation of the blood capillaries. The muscle fibres themselves do not increase in size very much, but there is an increase in the quantities of respiratory enzymes in them. There is also an increase in the amount of connective tissue in the muscles, so that they become less susceptible to minor injuries.

Disuse, such as occurs when a person is confined to bed or subject to prolonged weightlessness in a space station, leads to reduction of the muscle mass and especially of the contractile proteins. Hence the need for specific exercise regimes for invalids and astronauts.

CHAPTER TEN

The mechanism of contraction in skeletal muscle

In the previous chapter we examined some of the properties of skeletal muscles without giving much consideration to the mechanisms of the contraction process. It is as if we had investigated the properties of a motor vehicle by measuring its top speed, its fuel consumption, and so on, without finding out how the engine works. Now it is time to look under the bonnet.

EXCITATION–CONTRACTION COUPLING

The way in which the muscle cell is excited has been described in Chapter 7: an all-or-nothing action potential sweeps along the whole length of the fibre. This is followed by contraction, and the process linking the two events is called the excitation–contraction coupling process. The question we have to answer is: how does the action potential cause contraction?

Depolarization of the cell membrane

When muscle fibres are immersed in a solution containing a high concentration of potassium ions, they undergo a relatively prolonged contraction called a *potassium contracture*. The tension produced is related to the potassium concentration in a sigmoidal way as it shown in Fig. 10.1. The membrane potential is of course reduced under these conditions (see Chapter 3), so it seems that depolarization is an adequate stimulus for contraction. Normally this depolarization occurs during the propagated action potential.

Contractures can also be produced by various drugs which depolarize the cell membrane, such as acetylcholine, veratridine and others. Caffeine also causes contractures, but without producing much depolarization; it is apparently affecting a later stage in the coupling process.

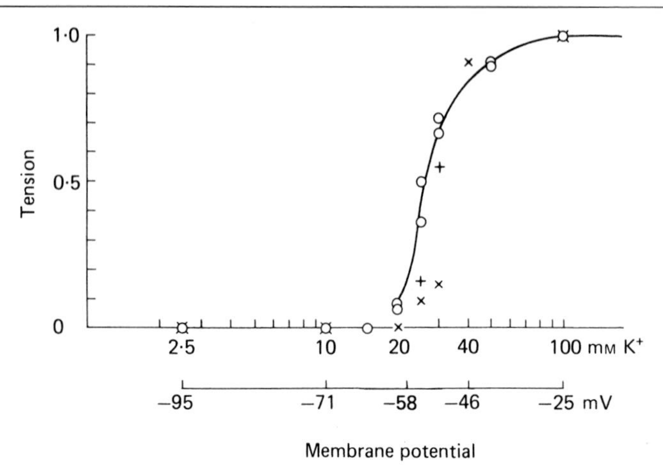

Fig. 10.1. The relation between peak contracture tension and potassium ion concentrations or membrane potential in single frog muscle fibres. From Hodgkin and Horowicz (1960).

Calcium ions

Myofibrils can be isolated from muscle by homogenizing the cells followed by differential centrifugation of the homogenate. Such a myofibrillar fraction will split ATP (just as occurs in the contracting muscle, we may assume), but only in the presence of calcium ions. The concentration of calcium ions required is about 10^{-6} M. This level is low but not negligible; it is higher than the free calcium ion concentration in the sarcoplasm of the resting muscle.

This observation suggests a way in which muscular contraction might be controlled: depolarization might cause an increase in the intracellular calcium ion concentration which would then activate the contractile apparatus. Direct evidence for this idea has been provided by C. C. Ashley and E. B. Ridgeway. They used the protein aequorin, isolated from a bioluminescent jellyfish, which emits light in the presence of calcium ions. Solutions of aequorin were injected into the large muscle fibres of the marriageable barnacle, *Balanus nubilis*. When such a fibre was stimulated electrically it produced a faint glow of light, indicating the presence of free calcium ions in its interior. The light output could be measured by using a photomultiplier tube, with the results shown in

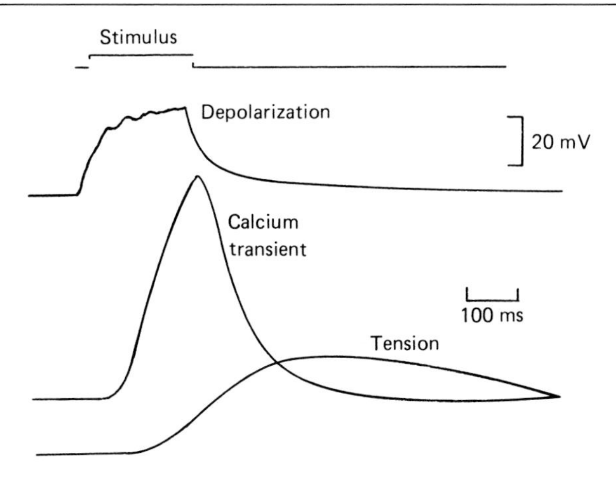

Fig. 10.2. Calcium transient in a barnacle muscle fibre, measured by the aequorin technique. The top trace monitors the stimulus pulse, and the second trace shows the ensuing depolarization. The third trace shows the photomultiplier output, indicating the concentration of free calcium ions inside the muscle cell. Finally the bottom trace shows the tension developed. After Ashley and Ridgeway (1968), redrawn.

Fig. 10.2. Notice that the time course of the 'calcium transient' is a little slower than that of the depolarization but much faster than that of the ensuing tension change.

Internal membrane systems

The next question to arise is, how does excitation at the cell surface cause release of calcium ions inside the fibre? The first step in the solution of this problem was the demonstration by A. F. Huxley and R. E. Taylor that there is a specific inward-conducting mechanism located at the Z line in frog sartorius muscles. (We shall examine the striation pattern in detail later. But here it is worth noting that the markedly birefringent A bands alternate with the less birefringent I bands, and that a thin dense line, the Z line, bisects the I bands. Fig. 10.6 shows the whole pattern.)

The muscle fibres were viewed by polarized light microscopy so as to make the striation pattern visible. They were stimulated by passing depolarizing current through an external microelectrode applied to the fibre surface. The stimulus was effective only when

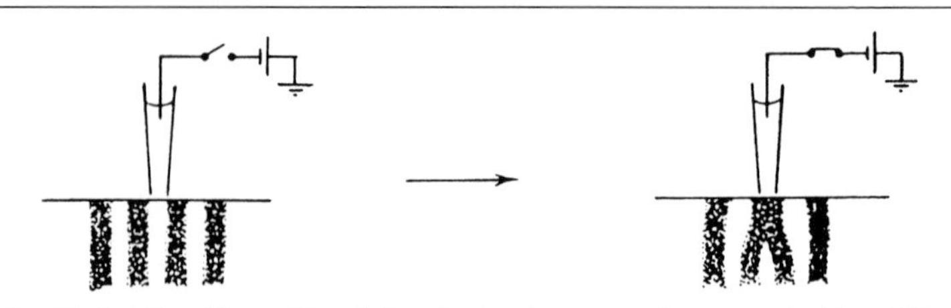

Fig. 10.3. The effect of local depolarizations on a frog muscle fibre. When the electrode is opposite certain active spots in the I band, as shown here, contraction ensues. Based on Huxley and Taylor (1958).

the electrode was positioned at certain 'active spots' located on the fibre surface in rows opposite the Z lines. In these cases the A bands adjacent to the I band opposite the electrode were drawn together, as is shown in Fig. 10.3.

At first it was thought that the inward-conducting mechanism was the Z line itself, but on repeating the experiments with crab muscle fibres it was found that the 'active spots' were localized not at the Z line but near the boundary between the A and I bands. This suggests that there is some transverse structure located at the Z lines in frog muscles and at the A–I boundary in crab muscles.

Such a structure was found in the electron microscopic examination of the *sarcoplasmic reticulum* in various skeletal muscles. This consists of a network of vesicular elements surrounding the myofibrils (Fig. 10.4). At the Z lines in frog muscles, and at the A–I boundaries in most other striated muscles (including crab muscles), are structures known as 'triads', in which a central tubular element is situated between two vesicular elements. These central elements of the triads are in fact tubules which run transversely across the fibre and are known as the transverse tubular system or *T system*. There is no continuity between the insides of the tubules of the T system and the vesicles of the sarcoplasmic reticulum, although their membranes are in close contact. The T system tubules are invaginations of the cell surface membrane. They open to the exterior at a limited number of sites corresponding to Huxley and Taylor's 'active spots'.

The sarcoplasmic reticulum consists of a series of membrane-bound sacs between the myofibrils. Vesicles formed from these sacs

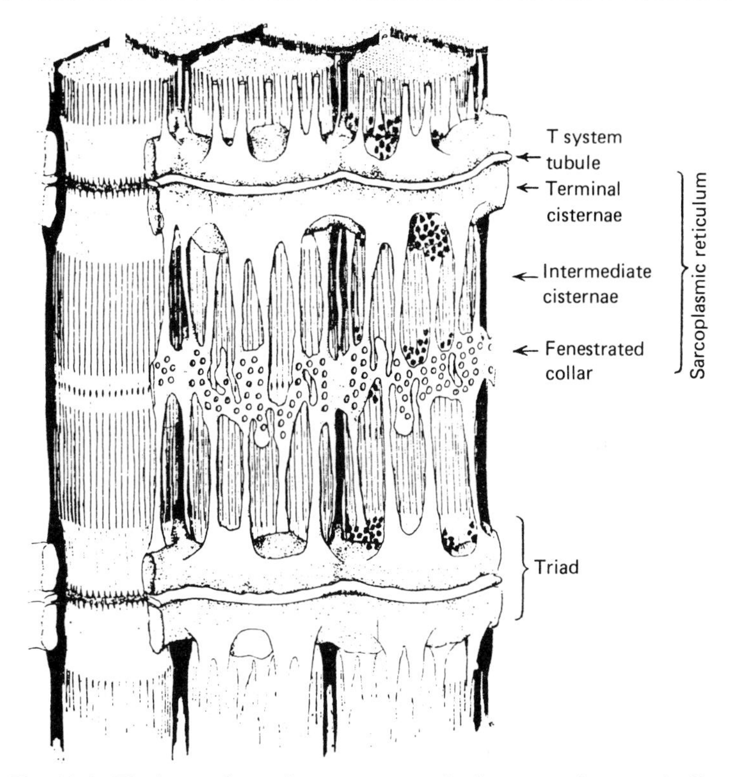

Fig. 10.4. The internal membrane systems of a frog sartorius muscle fibre. From Peachey (1965).

can be isolated from homogenised muscle by differential centrifugation. Their most interesting property is that they will accumulate calcium ions against a concentration gradient, by means of a 'calcium pump' which requires energy from the splitting of ATP for its activity. It seems very likely that the calcium ion concentration in the sarcoplasm of the living muscle is maintained at its low resting level by the action of this calcium pump. Calcium uptake by the sarcoplasmic reticulum could also account for the relaxation of the muscle after a contraction.

From circumstantial evidence it is possible to build up a convincing story of what probably happens during the coupling process (Fig. 10.5). We know that depolarization spreads down the

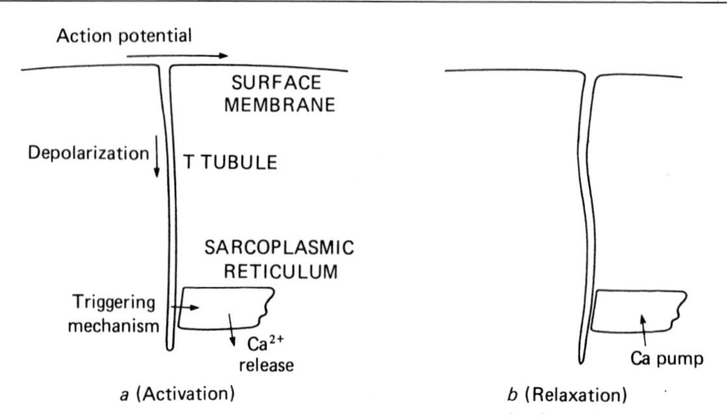

Fig. 10.5. Schematic diagram of the coupling process in skeletal muscle. During activation calcium ions are released from the sarcoplasmic reticulum (*a*). They are then pumped back into the sarcoplasmic reticulum, so causing relaxation (*b*).

T tubule into the interior of the fibre. We know that the sarcoplasmic reticulum is intimately linked to the T system at the triads and that it contains a considerable amount of calcium. It seems very reasonable to suggest that depolarization of the T tubule produces a change in the permeability of the sarcoplasmic reticulum membrane so as to release calcium ions into the myofibrils. Just how the link between T tubule and sarcoplasmic reticulum works is much less certain; one possibility is that membrane charge movements, analogous to the gating currents of the axon membrane, are involved. The released calcium then activates the contractile mechanism and the muscle contracts. Then calcium is pumped back into the sarcoplasmic reticulum and the muscle relaxes.

How does the calcium activate the contractile mechanism? Before answering that question we must first consider how the contractile mechanism itself works.

THE STRUCTURE OF THE MYOFIBRIL

Each muscle fibre contains a large number of thin longitudinal elements called myofibrils. They have characteristic banding patterns, and the bands on adjacent myofibrils are transversely

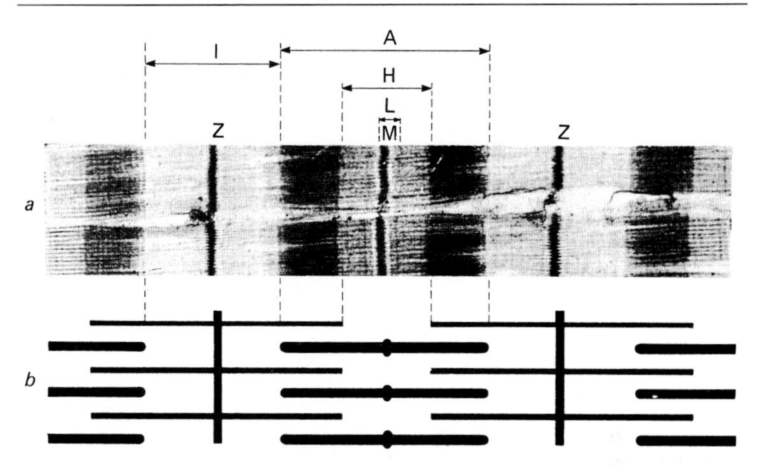

Fig. 10.6. The striation pattern of a vertebrate skeletal muscle fibre as seen by electron microscopy of thin sections (*a*), and its interpretation as two sets of interdigitating filaments (*b*). Photograph for *a* supplied by Dr H. E. Huxley.

aligned so that the whole fibre appears striated. In order to see the striations by light microscopy it is necessary to fix and stain the fibres, or to use phase-contrast, polarized light or interference microscopy.

Fig. 10.6*a* shows the striation pattern as seen by these methods and by low-power electron microscopy. The two main bands are the dark, strongly birefringent A band and the lighter, less birefringent I band. These bands alternate along the length of the myofibril. In the middle of each I band is a dark line, the Z line. In the middle of the A band is a lighter region, the H zone, which is bisected by a darker line, the M line. A lighter region in the middle of the H zone, the L zone, can sometimes be distinguished. (The letters used to describe the striation pattern are mostly the initials of names which are now no longer used.) The unit of length between two Z lines is called the *sarcomere*.

In the 1950s the use of the techniques of electron microscopy and thin sectioning, especially by H. E. Huxley and his colleagues, enabled the structural basis of the striation pattern to be discerned (Figs 10.6*b*, 10.7). The myofibrils are composed of two sets of filaments, thick ones about 11 nm in diameter and thin ones about

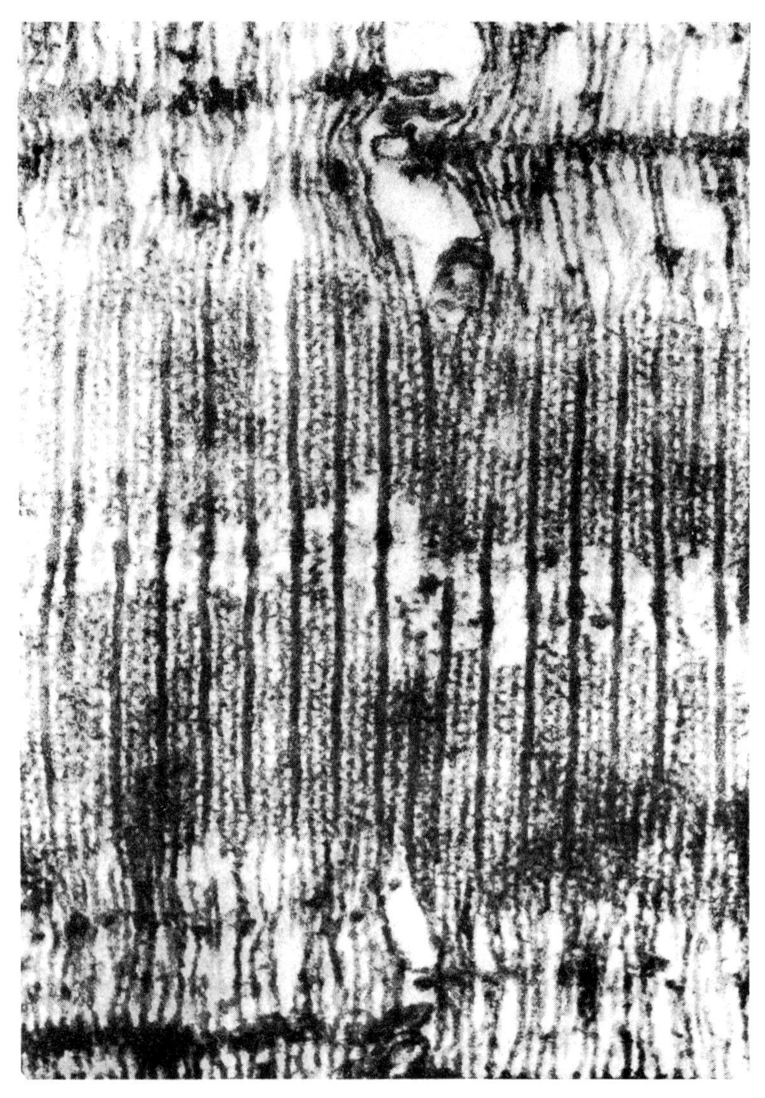

Fig. 10.7. Thin longitudinal section of a glycerol-extracted rabbit psoas muscle fibre. Notice, particularly, the cross-bridges between the thick and thin filaments. Photograph supplied by Dr H. E. Huxley.

5 nm in diameter. The thick filaments run the length of the A band. The thin filaments are attached to the Z lines and extend through the I bands into the A bands. The H zone is the region of the A band between the ends of the two sets of thin filaments, and the M line is caused by cross-links between the thick filaments in the middle of the sarcomere. The thick filaments have projections from them except in a short central region which corresponds to the L zone. In the overlap region these projections may be attached to the thin filaments so as to form *cross-bridges* between the two sets of filaments.

The myofibrils contain two major proteins, *actin* and *myosin*, and a number of minor ones. When myofibrils are washed in a solution which dissolves myosin the A bands disappear, and further washing with a solution which dissolves actin causes the I bands to disappear. This suggests that the thick filaments are composed largely of myosin and the thin filaments largely of actin.

THE SLIDING FILAMENT THEORY

Prior to 1954, most suggestions as to the mechanism of muscular contraction involved the coiling and contraction of long protein molecules, rather like the shortening of a helical spring. In that year the *sliding filament theory* was independently formulated by H. E. Huxley and Jean Hanson (using phase-contrast microscopy of myofibrils from glycerol-extracted muscles) and by A. F. Huxley and R. Niedergerke (using interference microscopy of living muscle fibres). In each case the authors showed that the A band does not change in length either when the muscle is stretched or when it shortens actively or passively. This suggests that contraction is brought about by movement of the thin filaments between the thick filaments, as is shown in Fig. 10.8. The sliding is thought to be caused by a series of cyclic reactions between the projections on the myosin filaments and active sites on the actin filaments; each projection first attaches itself to the actin filament to form a cross-bridge, then pulls on it and finally releases it, moving back to attach to another site further along the actin filament.

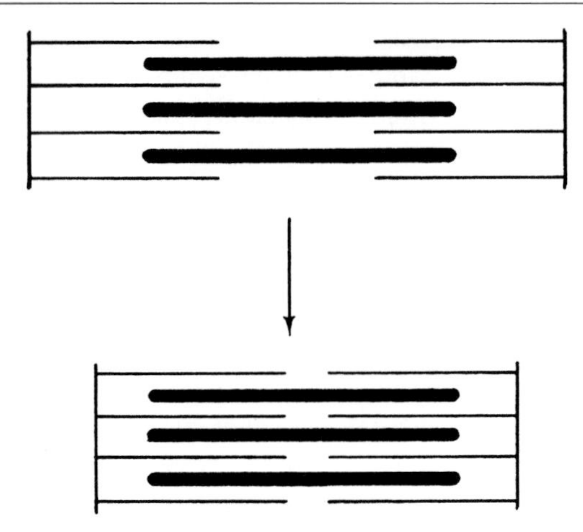

Fig. 10.8. The structural changes in a sarcomere on shortening, according to the sliding filament theory.

Let us now have a look at some of the further evidence for this theory.

The lengths of the filaments

Electron microscopy provides evidence that the filaments do not change in length when the muscle is stretched or allowed to shorten. This is just what we would expect if such length changes involve sliding of the filaments past each other. Measurements on frog muscles suggest that the thick filaments are 1.6 μm long, and that the thin filaments extend for 1.0 μm on each side of the Z line.

Electron microscopy has to be performed on muscle tissue which has been fixed and stained. There is a need, therefore, for some length-measuring method which can be applied directly to the living muscle. This is provided by the technique of X-ray diffraction, in which a beam of X-rays is passed through a muscle and the resulting diffraction pattern indicates the distances between repeat units in the muscle structure.

In the thick filaments the X-ray diffraction pattern suggests that there are structures which repeat axially at 14.3 nm and helically at 42.9 nm. In the thin filaments there seem to be structures arranged on helices with pitches of 5.1, 5.9 and about 37 nm.

The structural basis of these repeat distances we shall return to later. What is important in our present context is that they do not change when the muscle is lengthened or shortened, either when it is resting or during contraction. This provides further evidence that the filaments themselves do not shorten or lengthen during the corresponding changes in muscle length.

The relation between sarcomere length and isometric tension

The suggestion that contraction depends on the interaction of the actin and myosin filaments at the cross-bridges implies that the isometric tension should be proportional to the degree of overlap of the filaments. In order to test this idea it is necessary to measure the active increment of isometric tension at different known sarcomere lengths. The measurements have to be done on a single fibre and there are some technical difficulties because sarcomeres at the ends of a fibre may take up lengths different from those in the middle. A. F. Huxley and his colleagues overcame these difficulties by building an apparatus which used optical servomechanisms to maintain the sarcomere lengths in the middle of a fibre constant during a contraction.

Fig. 10.9 summarizes the results of these experiments. It is evident that the length–tension diagram consists of a series of straight lines connected by short curved region. There is a 'plateau' of constant tension at sarcomere lengths between 2.05 and 2.2 μm. Above this

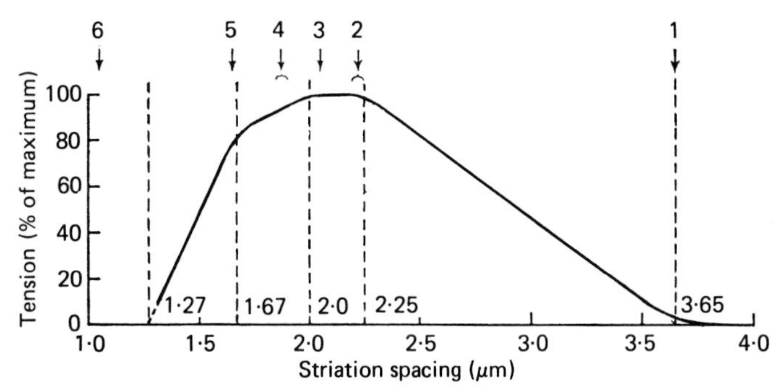

Fig. 10.9. The isometric tension (active increment) of a frog muscle fibre at different sarcomere lengths. The numbers 1 to 6 refer to the myofilament positions shown in Fig. 10.11. From Gordon, Huxley and Julian (1966).

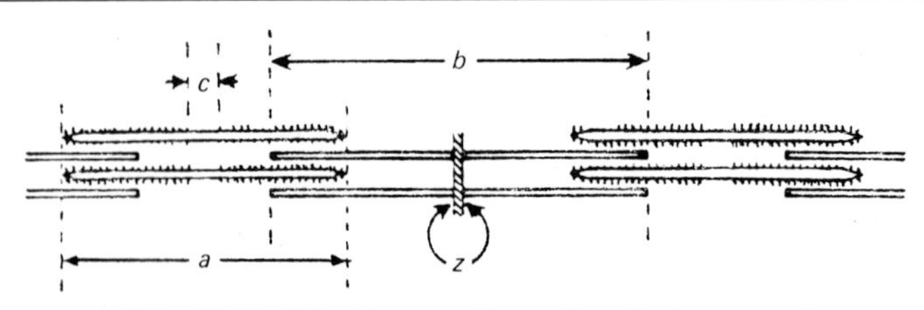

Fig. 10.10. Myofilament dimensions in frog muscle.

range tension falls linearly with increasing length; the projected line through most of the points in this region reaches zero at 3.65 μm. Below the plateau, tension falls gradually with decreasing length down to about 1.65 μm, then must more steeply, reach zero at about 1.3 μm.

Does this curve fit the predictions of the sliding filament theory? We need to know the dimensions of the filaments (Fig. 10.10). Measurements by electron microscopy indicate that the myosin filaments are 1.6 μm long and the actin filaments, including the Z line, are 2.05 μm long. The middle region of the myosin filaments, which is bare of projections and therefore cannot form cross-bridges, is 0.15 to 0.2 μm long and the thickness of the Z line is about 0.05 μm.

Now let us see if the length–tension diagram shown in Fig. 10.9 can be related to these dimensions, starting at long sarcomere lengths and working through to short ones. Above 3.65 μm (stage 1 in Fig. 10.11) there should be no cross-bridges formed, and therefore no tension development. In fact there is some tension development up to about 3.8 μm; this might well be due to some residual irregularities in the system. Between 3.65 μm and 2.2 to 2.25 μm (1 to 2) the number of cross-bridges increases linearly with decrease in length, and therefore the isometric tension should show a similar increase. It does. With further shortening (2 to 3) the number of cross-bridges remains constant and therefore there should be a plateau of constant tension in this region. There is. After stage 3 we might expect there to be come increase in the internal resistance to shortening since the actin filaments must how overlap, and after stage 4 the actin filaments from one half of the sarcomere might interfere with the cross-bridge formation in the

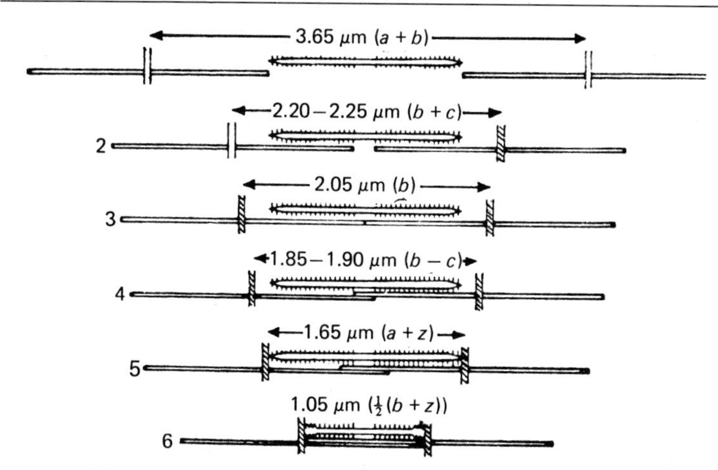

Fig. 10.11. Myofilament arrangements at different lengths. The letters a, b, c and z refer to the dimensions given in Fig. 10.10. From Gordon, Huxley and Julian (1966).

other half of the sarcomere. We would expect both these effects to reduce the isometric tension, which does indeed fall at lengths below 2.0 μm. At 1.65 μm (stage 5) the myosin filaments will hit the Z line, and so there should be a considerable increase in the resistance to further shortening; there is a distinct kink in the curve at almost exactly this point, after which the tension falls much more sharply. The curve reaches zero tension at about 1.3 μm, before stage 6 is reached.

It would be difficult to find a more precise test of the sliding filament theory than is given by this experiment, and the theory clearly passes the test with flying colours.

THE MOLECULAR BASIS OF CONTRACTION

The myofibrils consist of a small number of different proteins, of which *actin* and *myosin* are the most important and are involved with the splitting of ATP and the process of contraction. *Tropomyosin* and the *troponins* are concerned with the control of this actin–myosin interaction. There are also a number of other minor proteins whose functions are for the most part less evident.

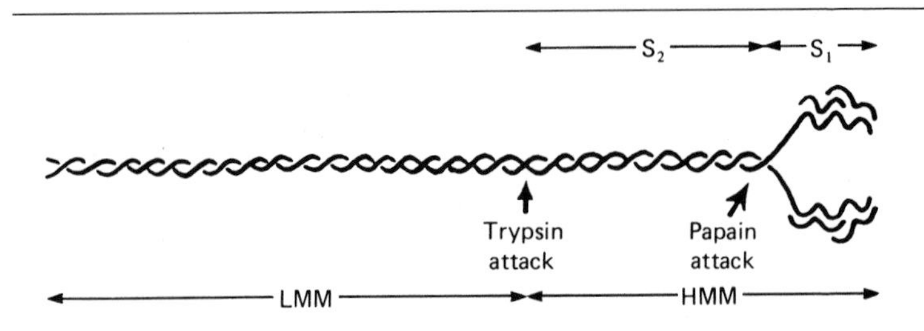

Fig. 10.12. Schematic diagram of the myosin molecule. Binding sites for actin and ATP occur in the S_1 subunits.

Myosin

Myosin (Fig. 10.12) is a rather complex protein with a molecular weight of about 470000. One of its most important properties is that it is an ATPase, i.e. it will enzymatically hydrolyse ATP to form ADP and inorganic phosphate. Treatment with the proteolytic enzyme trypsin splits the myosin molecule into two sections, known as light meromyosin and heavy meromyosin; only heavy meromyosin acts as an ATPase. Examination of the meromyosins by electron microscopy shows that heavy meromyosin has a more or less globular 'head' with a short 'tail', whereas light meromyosin is a rod-like molecule. Heavy meromyosin can be further split by digestion by papain, to give two globular S_1 subfragments and a rod-like S_2 subfragment. The ATPase activity is confined to the S_1 subfragment. Light meromyosin molecules will aggregate to form filaments under suitable conditions, but neither heavy meromyosin nor its two subfragments will.

H. E. Huxley found that under the right conditions myosin molecules can aggregate to form filaments. These filaments had regularly spaced projections on them which almost certain correspond to the projections and cross-bridges seen in thin sections of myofibrils. In the middle of each filament was a section from which these projections were absent, which must correspond to the L zone of intact muscle fibres. Similar filaments could be isolated from homogenised myofibrils; they were all 1.6 μm long, whereas the 'artificial' filaments were variable in length. Huxley suggested that the 'tails' of the myosin molecules become attached to each other to form a filament as shown in Fig. 10.13, with the 'heads' projecting

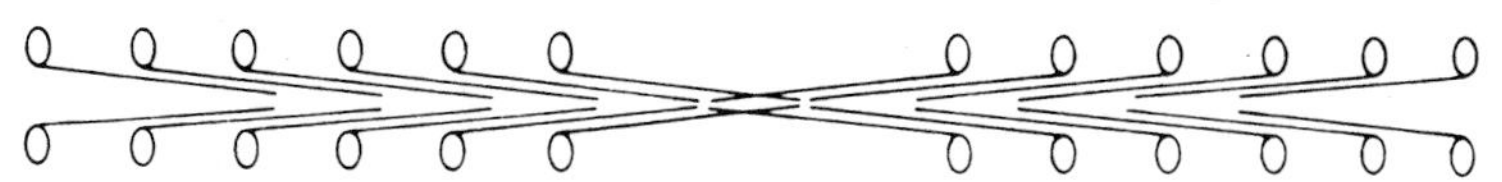

Fig. 10.13. H. E. Huxley's suggestion as to how the myosin molecules aggregate to form a thick filament with a projection-free shaft in the middle and reversed polarity of the molecules in each half of the sarcomere. From Huxley (1971).

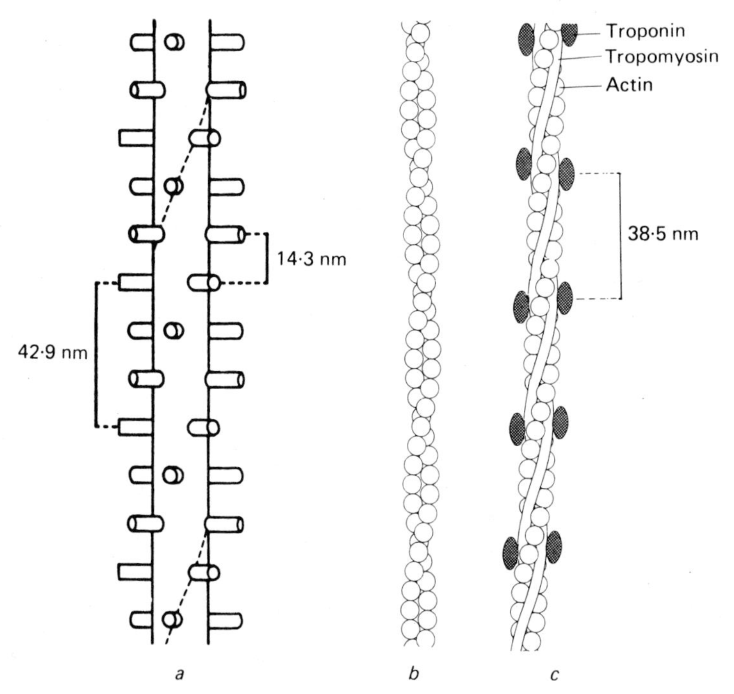

Fig. 10.14. Models of the structure of the thick and thin filaments. *a*, the myosin filament; *b*, F-actin; *c*, the thin filament. Based on Offer (1974) after various authors.

from the body of the filament. Notice particularly that this arrangement accounts for the bare region in the middle, and also that it implies that the polarity of the myosin molecules is reversed in the two halves of the filament.

X-ray diffraction studies show that there is an axial repeat of 14.3 nm and a helical repeat of 42.9 nm on the myosin filament, as has already been mentioned. This suggests that a group of myosin heads emerges from the filament every 14.3 nm, and that their orientation rotates so that every third group is in line. There are two, three or perhaps four myosin molecules in each group. A model for three molecules per group is shown in Fig. 10.14*a*; this is perhaps the most probable arrangement.

Actin

Isolated actin exists in two forms: G-actin, a more or less globular molecule of molecular weight about 42000, and F-actin, a fibrous protein which is a polymer of G-actin. Neither form has any ATPase activity, but they will both combine with myosin.

F-actin consists of two chains of monomers connected together in the form of a double helix, as is shown in Fig. 10.14*b*. The thin filaments in intact muscle also contain tropomyosin and troponin, probably arranged as in Fig. 10.14*c*.

The interaction of actin, myosin and ATP

If solutions of actin and myosin are mixed, a great increase in viscosity occurs, due to the formation of a complex called *actomyosin*. Actomyosin is an ATPase which is activated by magnesium ions. 'Pure' actomyosin (a mixture of purified actin and purified myosin) will split ATP in the absence of calcium ions. However 'natural' actomyosin (an actomyosin-like complex which can be extracted from minced muscle with strong salt solutions, and which also contains tropomyosin and troponin) can only split ATP if there is a low concentration of calcium ions present. In the absence of calcium ions, addition of ATP to a solution of natural actomyosin results in a decrease in viscosity, suggesting that the actin–myosin complex becomes dissociated.

We can use these observations to make plausible suggestions about how actin and myosin interact within the filament array in which they exist in the living muscle. In the resting condition there is an adequate concentration of ATP and a very low concentration of calcium ions, so there is no interaction between the actin and myosin and no ATP splitting. On activation the calcium ion concentration rises and so cross-bridges are formed between the two sets of filaments, ATP is split and sliding occurs.

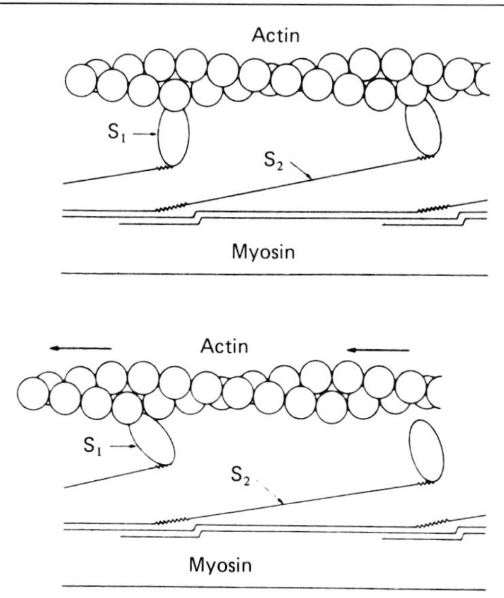

Fig. 10.15. Showing how sliding is probably brought about, by a rotation of the S_1 subunits about their points of attachment to the thin filament. In the upper diagram the left-hand cross-bridge has just attached, whereas the S_1 subunit of the right-hand one has nearly completed its rotation. The lower diagram shows the situation a short time later: the S_1 subunit of the left-hand cross-bridge has rotated, so pulling the actin filament to the left, and the right-hand cross-bridge is now detached. From Huxley (1976).

It seems likely that the myosin molecule possesses two flexible linkages, one between S_1 and S_2 and the other between S_2 and light meromyosin. The primary source of movement is probably a rotation in the S_1 head near its point of attachment to the actin filament (Fig. 10.15). Each cross-bridge therefore undergoes a cycle of events whereby it attaches to the actin filament, rotates (causing sliding of the actin and myosin filaments relative to each other) and then detaches. The cross-bridge is then ready to attach to a new site on the actin and repeat the cycle. The energy for each turn of the cycle is provided by the breakdown of one molecule of ATP to ADP and inorganic phosphate (Fig. 10.16).

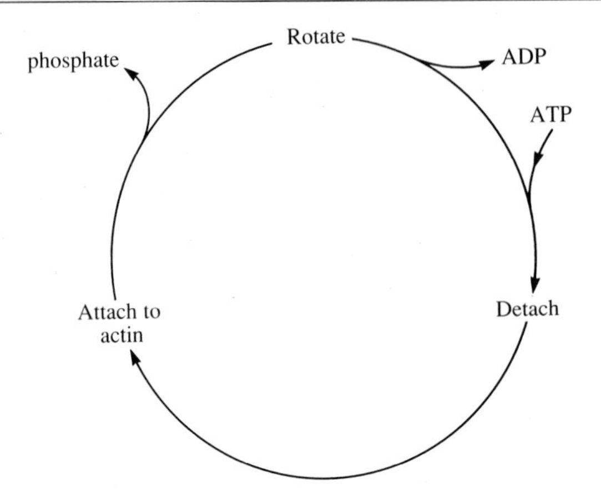

Fig. 10.16. The cycle of cross-bridge activity during contraction.

The molecular basis of activation

As mentioned earlier, pure actin will react with pure myosin so as to split ATP in the absence of calcium ions. But if tropomyosin and troponin are also present, the actin–myosin interaction and ATP splitting will occur in the presence of calcium ions. Hence it is probable that tropomyosin and troponin are intimately involved in the control of muscular contraction.

Tropomyosin is a fibrous protein which will bind to actin and troponin. Troponin is a globular protein with three subunits: one binds to actin, another to tropomyosin and a third combines reversibly with calcium ions, undergoing a conformational change in the process.

The molecular ratios of actin, tropomyosin and troponin in the muscle are 7:1:1. A model of the thin filament incorporating these ratios is shown in Fig. 10.14*c*, where the tropomyosin molecules lie in the grooves between the two chains of actin monomers and a troponin molecule is attached with each tropomyosin molecule to every seventh actin monomer. This arrangement would give a repeat distance of $(7 \times 5.5) = 38.5$ nm for the troponin and tropomyosin, which agrees well with the observation of X-ray reflections at this distance.

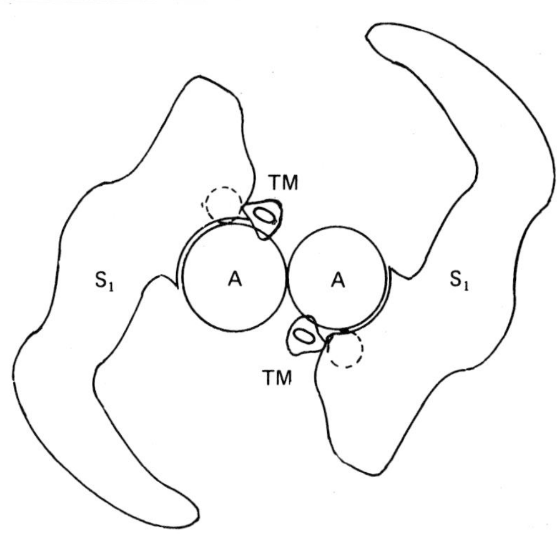

Fig. 10.17. One of the models proposed to show how movement of tropomyosin molecules may affect actin–myosin interactions. A thin filament is seen in cross-section with actin (A) and tropomyosin (TM) molecules. Two myosin S_1 subunits are shown attached to the thin filaments. Tropomyosin positions are shown for the muscle at rest (dotted circle) and when active (contours). Other models give somewhat different shapes and positions for the various protein molecules, but all agree that the tropomyosin molecule moves into the 'groove' between the actin monomers on activation. It is thought that the S_1 heads are unable to attach to the actin filament until this movement takes place. From Huxley (1976).

Evidence about the structure of the thin filament under different conditions has been obtained from X-ray diffraction measurements and from computer analysis simulating optical diffraction of electron micrographs. It now seems likely that the binding of calcium by a troponin molecule causes a change in its shape which draws the tropomyosin molecule to which it is attached further into the groove between the two chains of actin monomers (Fig. 10.17). In the resting condition it looks as though the tropomyosin molecules prevent actin–myosin interaction by covering the myosin binding sites on the actin monomers. So this movement on activation has the exciting consequence that each tropomyosin molecule uncovers the myosin-binding sites on seven actin monomers. The myosin heads can then combine with the actin and so the muscle contracts. It is a very elegant piece of biological machinery.

CHAPTER ELEVEN

Non-skeletal muscles

Muscle cells have become adapted to a variety of different functions during their evolution, so that the details of the contractile process and its control are not always identical with those in vertebrate skeletal muscles. In this chapter we examine the properties of mammalian heart and smooth muscles.

CARDIAC MUSCLE

Mammalian heart muscle consists of a large number of branching uninucleate cells connected to each other at their ends by *intercalated discs* (Fig. 11.1). Electron micrography shows that the intercalated discs consist largely of accumulations of dense material on the insides of the two cell membranes; these apparently serve to fix the cells together and allow the filaments of the contractile apparatus in one cell to pull on those of the next one in the line. Gap junctions are also present in the intercalated discs and these allow electrical currents to flow from one cell to another.

The contractile apparatus is much the same as in skeletal muscles, with thick myosin and thin actin filaments aligned transversely so that the muscle cells as a whole are cross-striated in appearance. As

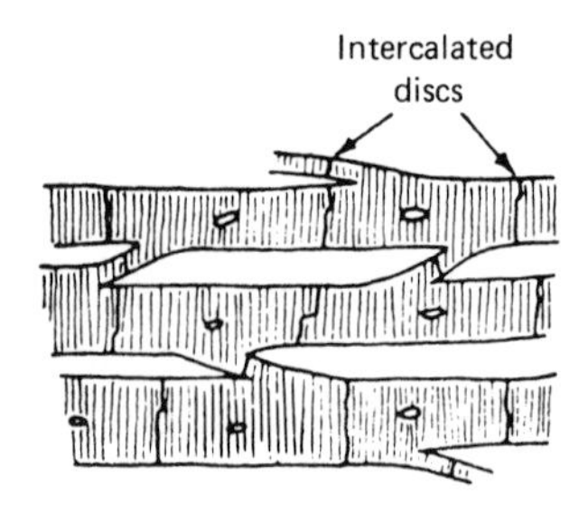

Fig. 11.1. Mammalian heart muscle cells.

in a skeletal muscle fibre, the interior of the cell also contains mitochondria, sarcoplasmic reticulum and the transverse tubules of the T system.

The cardiac action potential

Intracellular recordings from heart muscle fibres were first made using isolated bundles of Purkinje fibres from dogs. The Purkinje fibres form a specialized conducting system which serves to carry excitation through the ventricle. After being isolated for a short time, they begin to produce rhythmic spontaneous action potentials, of the sort shown diagrammatically in Fig. 11.2. The form of these action potentials differs from those of nerve axons and twitch skeletal muscle fibres in that there is a prolonged 'plateau' between the peak of the action potential and the repolarization phase. The action potential is preceded by a slowly-rising *pacemaker potential*, which acts as a trigger for the action potential when it crosses a threshold level.

What is the ionic basis of these heart muscle action potentials? The peak membrane potential is reduced when the external sodium ion concentration is lowered. This suggests that the initial rapid depolarization is brought about by a regenerative increase in the sodium conductance of the membrane, just as in the action potential

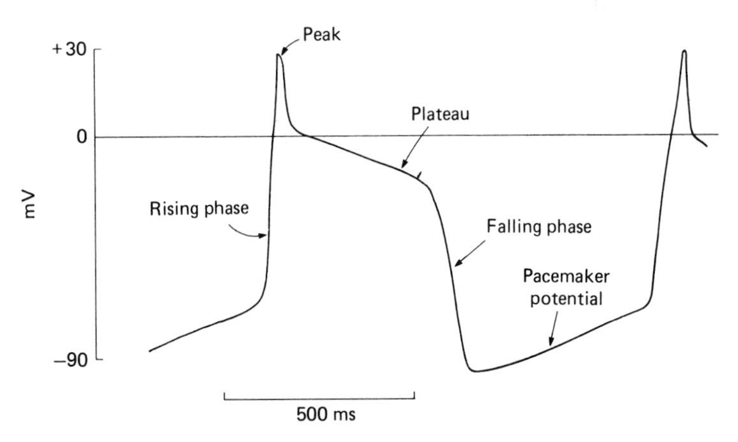

Fig. 11.2. The cardiac action potential. Based on the action potentials produced spontaneously in isolated Purkinje fibres.

of nerve axons. However, cardiac action potentials must involve phenomena which are absent from nerve axons, in order to explain the plateau and the pacemaker potential.

Sorting out the full nature of the cardiac action potential has proved to be a complicated and as yet unfinished task (Noble, 1984). In comparison with squid axons, the size and geometry of heart muscle fibres makes it much harder to subject them to voltage-clamp, and the number of different ionic channels involved in the action potential is larger. A computer model, based on voltage-clamp measurements on Purkinje fibres, has been produced by D. DiFrancesco and D. Noble; let us have a brief look at it.

There are four ionic conductances involved in the model. The sodium conductance g_{Na} is rapidly activated and then inactivated by depolarization, and blocked by tetrodotoxin; it is essentially similar to the sodium conductance of nerve and skeletal muscle cells. The potassium conductance g_K is complex (at least three different types of channel seem to be involved), with some components being activated by hyperpolarization and others by depolarization. There is an appreciable calcium conductance g_{Ca} which is activated by depolarization and produces inward current during the plateau. A fourth conductance g_f permits the slow inward movement of sodium and other ions; it is activated by hyperpolarization and is important during the pacemaker potential.

The sequence of events in the DiFrancesco–Noble model is shown in Fig. 11.3. Let us begin at the point in the cycle where the membrane potential is at its most negative, at about 0.4 ms on the time trace. It has reached this negative value because the potassium conductance g_K is high. However, the pacemaker conductance g_f has been switched on by the hyperpolarization, and it rises steadily for the next second or so. The slow sodium ion inflow which this permits results in a steady depolarization, the pacemaker potential. After a time the pacemaker potential has depolarized the membrane sufficiently to open the fast-activating sodium channels. Then follows the familiar runaway relation between membrane potential and sodium conductance just as in the nerve axon, so that there is a massive inflow of sodium ions and a rapid overshooting depolarization. This increase in sodium conductance is rapidly

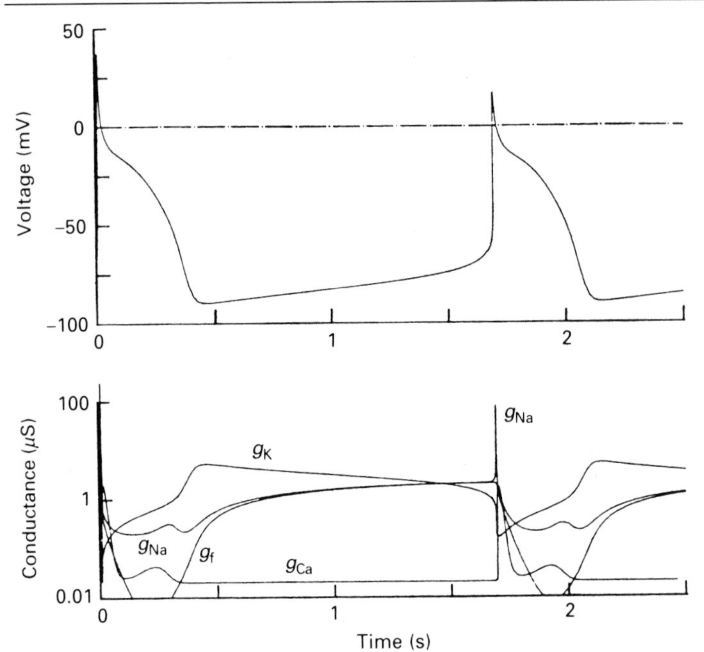

Fig. 11.3. Computer simulation of the cardiac action potential. The associated conductance changes are shown in the lower graphs: g_f is the inward current which becomes apparent during the pacemaker potential. The sodium conductance g_{Na} includes both the conductance due to fast sodium channels and the sodium component of g_f. From DiFrancesco and Noble (1985).

inactivated (the sodium channels close again) so that the membrane potential rapidly falls back from its positive peak, just as in the nerve action potential.

But now the model departs radically from the situation in nerve axons. The potassium conductance has fallen to a low level and there is an elevated calcium conductance, hence the membrane potential remains near zero for some hundreds of milliseconds. This plateau declines gradually and is brought to an end as a result of the long-delayed increase in potassium conductance, and any calcium and fast sodium channels remaining open are finally closed during the repolarization phase. By the end of the action potential the pacemaker conductance g_f has already begun to rise and so the new cycle continues on its way.

The heart muscle cell membrane is refractory for a long time – some hundreds of milliseconds – after the completion of an action potential. Consequently it is not possible to tetanize heart muscle by repetitive stimulation, since the refractory period is long enough to allow the muscle to relax after each action potential. This is of vital importance to the functioning of the heart as a pump: the relaxation phase allows the heart to be refilled with blood from the veins before expelling it to the arteries during the contraction phase.

The long duration of the cardiac action potential as compared with that in a twitch skeletal muscle fibre is related to an important difference in their roles in excitation–contraction coupling. In the skeletal muscle the action potential acts simply as a trigger which initiates the resulting contraction but has no further control over it. But in the cardiac muscle the action potential is coincident with most of the contraction phase, and indeed relaxation begins during the repolarization phase (Fig. 11.4). If the action potential is shortened in some way, relaxation begins sooner and so the tension reaches a lower peak level; the reverse happens if the action potential is lengthened. Hence the action potential acts as a controller of the contraction as well as a trigger for it.

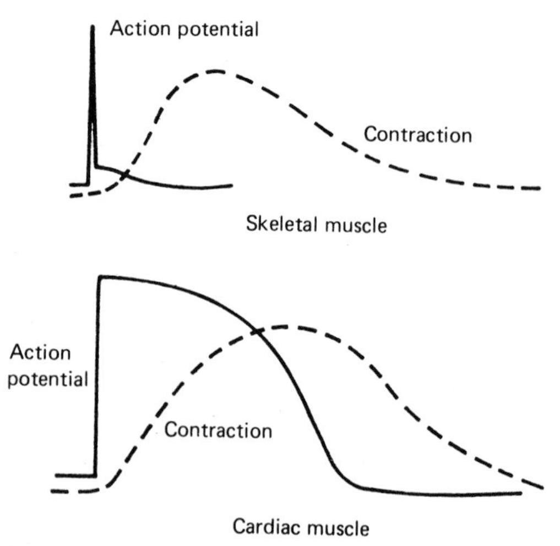

Fig. 11.4. Diagram comparing the relative time scales of the electrical and mechanical responses in skeletal and cardiac muscle. From Noble (1979).

The spread of excitation

A heart would be ineffective if the rhythmic action potentials and their associated contractions in any one fibre were independent of those of its fellows: it is obviously essential that the contractions of the fibres should be synchronized. The gap junctions in the intercalated discs are presumably the sites at which current can flow from one cell to the next. (The gap junctions are so named because the adjacent cell membranes are brought close together, leaving only a small gap between them. This gap is bridged by cross-connections which appear to be molecular pores through which, it is assumed, current can readily flow, see p. 120.) Hence local circuits set up by an action potential in one cell can cross into the next cell and so excite it.

The excitation sequence in mammals is modified by the existence of some cells, the Purkinje fibres, which have become specialized for conduction rather than contraction. They are longer and more elongated than other cardiac cells and contain little contractile material.

Excitation begins at the pacemaker region in the sinuatrial node (Fig. 11.5); the fibres here are spontaneously active and their

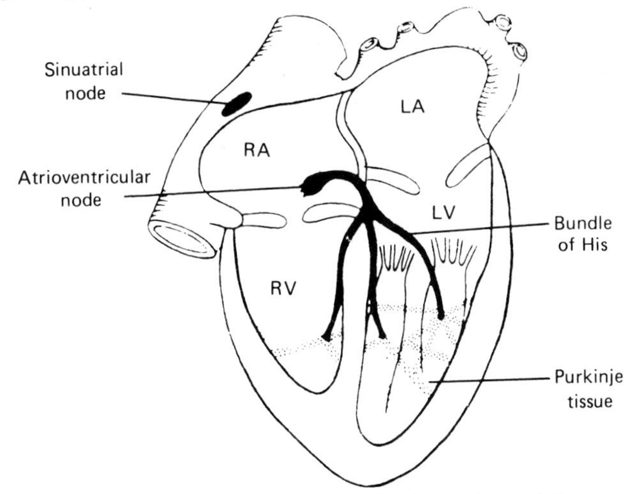

Fig. 11.5. Pacemaker and specialized conductile regions of the mammalian heart. From Scher (1965).

rhythmic production of action potentials determines the rate of beating of the whole heart. These then excite their neighbours by local current flow, and so a wave of depolarization spreads across the atria, resulting in a contraction of each atrium so that blood is forced into the ventricles. The atria are connected to the ventricles at the atrioventricular node. This consists of fibres which are small in diameter so that conduction at this point is slow; this ensures that there is an appreciable delay between the excitation of the atria and the ventricles. The atrioventricular node is connected to the large Purkinje fibres in the bundle of His. These serve to carry the excitation on to the main mass of the ventricular muscle, beginning in the septum and then spreading from the apex of the ventricle up to its base.

The electrocardiogram

It is possible to measure the electrical activity of the human heart simply by attaching leads to the wrists and ankles of the subject and connecting them to a suitable recording device. The resulting record is known as an electrocardiogram, or ECG for short. Conventional ECGs were first obtained by Einthoven, using the string galvanometer which he invented for the purpose. Their measurement is now usually done by means of a hot wire pen recorder, and has long been a standard procedure in medical practice. Fig. 11.6 shows a typical ECG, recorded between the right arm and the left leg. The

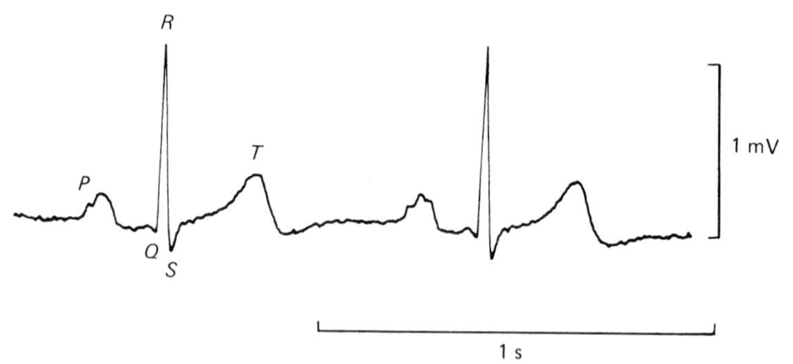

Fig. 11.6. A human electrocardiogram, recorded between electrodes applied to the right wrist and left ankle.

different peaks in the electrical cycle of events were labelled the *P*, *Q*, *R*, *S* and *T* waves by Einthoven. The events in the heart cycle to which these electrical waves correspond can be worked out by recording with surface electrodes from exposed hearts in experimental animals.

The *P* wave is produced by currents associated with the spread of excitation over the atria. During the plateau of the atrial action potentials there is little current flow and so the level of the ECG approximates to zero. The net currents involved in the excitation of the atrioventricular node and the specialized conducting tissue in the ventricle are small, because the number of cells involved is small, and so are not evident in the ECG. Then the depolarization of the large mass of ventricular cells is accompanied by large net currents which are seen as the *QRS* complex in the ECG. After this the whole of the ventricle is depolarized and there is very little electrical current flow. The ventricular muscle is contracting at this time to pump blood out along the aorta and pulmonary artery. Then repolarization of the ventricular fibres occurs, at slightly different times in different places, and the current flow associated with this is seen as the *T* wave. After this the heart is electrically at rest except in the pacemaker regions, its muscle cells are relaxing and it is refilling with blood ready for the next cycle. The pacemaker potentials preceding the next *P* wave are not visible in the ECG.

We can relate the form of the ECG to the time course of the action potentials in the cardiac muscle cells. If the action potentials were synchronous in all the muscle cells, there would be no current flow from one cell to another (because they would all be at the same potential at any one instant) and so there would be no external current flow and therefore no ECG. But since the excitation and repolarization move progressively across the heart there is an external current flow which is detected as the ECG. In other words the voltage recorded in the ECG at any moment is related to the differences in membrane potential of muscle cells in different regions of the heart at that moment. Hence we find that the rapidly changing currents constituting the *QRS* complex are related to the rapid depolarizations at the beginnings of the ventricular muscle cell action potentials, and the slower and smaller changes constituting the *T* wave correspond to their repolarization. The precise nature of these relations is geometrically rather complicated (see Scher, 1974), and it is not appropriate to examine the details here.

Nervous control of the heart

As we have seen, the heart beat originates as repetitive activity within the cells of the pacemaker regions in the heart. This activity can be modified by the action of nerve fibres innervating the heart via the autonomic nervous system. Stimulation of the vagus nerve, containing parasympathetic nerve fibres whose postganglionic terminals release acetylcholine, causes slowing of the heart rate. Stimulation of the sympathetic nerves innervating the heart, so releasing noradrenaline from their terminals, causes an increase in the heart rate and more powerful contractions of the heart muscle; similar effects are produced by the hormone adrenaline.

The inhibitory effects of acetylcholine seem to be brought about by an increase in the potassium conductance of the heart muscle cell membrane. The effects of adrenaline and noradrenaline are more complicated; they probably involve an increase in the calcium currents and a decrease in the potassium currents.

SMOOTH MUSCLE

Smooth muscles, otherwise known as unstriated or plain muscles, form the muscular walls of the viscera and blood vessels. They also occur in the iris, ciliary body and nictitating membrane in the eye, in the trachea, bronchi and bronchioles of the lungs, and are the small muscles which erect the hairs. Individual smooth muscle cells are uninucleate, and are much smaller than the multinucleate fibres of skeletal muscles. They are usually about 4 μm in diameter and up to 400 μm long. Adjacent cells are connected at regions where the opposing cell membranes are brought close together to form gap junctions; it seems very likely that current can flow from one cell to another at these sites.

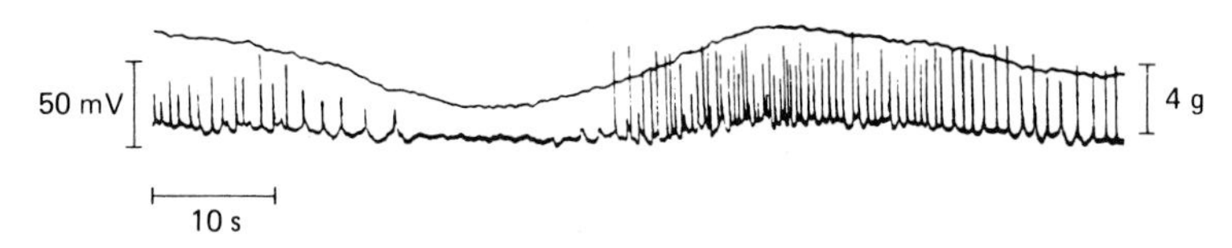

Fig. 11.7. Simultaneous records of tension (upper trace) and electrical activity in guinea-pig taenia coli. From Bülbring (1979).

Excitation

Many smooth muscles show a great deal of spontaneous activity. This is particularly so in intestinal muscles, where the spontaneous contractions serve to mix and move the gut contents. The electrical activity consists of slow waves of variable amplitude and all-or-nothing action potentials (Fig. 11.7). The fibres are depolarized and the frequency of action potentials increases if the muscle is stretched. The spontaneous activity can be modified by the action of extrinsic nerves, by adrenaline, and in uterine muscle by the action of hormones of the reproductive cycle. The smooth muscles of the iris, nictitating membrane and vas deferens are not spontaneously active.

The action potentials last for several milliseconds and are thus much closer than those of nerve axons and skeletal muscle cells. They are insensitive to tetrodotoxin and can often be produced in the absence of sodium ions, but are prevented by calcium channel blocking agents such as nifedipine. This suggests that calcium ions are the main carriers of inward current.

Action potentials can be initiated in sheets or strips of smooth muscle by electrical stimulation; they will then propagate along the axes of the muscle cells from one cell to another. There is also some slower propagation across the axes of the cells.

Only a proportion of the cells receive innervation from the nerves supplying the muscle. The electrical changes produced by nervous action on these cells spread to nearby cells by current flow from cell to cell, probably via the gap junctions by which they are connected. Stimulation of the nerves results in postsynaptic potentials of various types. Excitatory nerves produce depolarizing potentials whereas inhibitory nerves produce hyperpolarizing ones.

There appear to be at least three transmitter substances involved (see Burnstock, 1979). Acetylcholine is an excitatory transmitter in much intestinal muscle and in the iris. Noradrenaline is excitatory at some sites, as in the vas deferens, and inhibitory at others, as in intestinal muscle and the iris of the eye. Adenosine triphosphate is also an important inhibitory transmitter in intestinal muscle.

Contraction

Smooth muscles contain the major contractile proteins actin and myosin, together with tropomyosin. The relative proportion of myosin is much less than in skeletal muscles. The activation mechanism is calcium-dependent, but does not act via troponin, which is absent.

The actin occurs in thin filaments which are readily seen by electron microscopy. Many of them are attached to dense bodies in the cytoplasm, others are attached to dense patches next to the cell membrane. The structure of the thick filaments is different from that in skeletal muscle. The myosin molecules are probably oriented in opposite directions on the two faces of a filament. This arrangement allows a thin filament to be pulled over the whole of its length by a thick filament, so that the muscle can operate at near maximum tension over a wide range of lengths (Fig. 11.8).

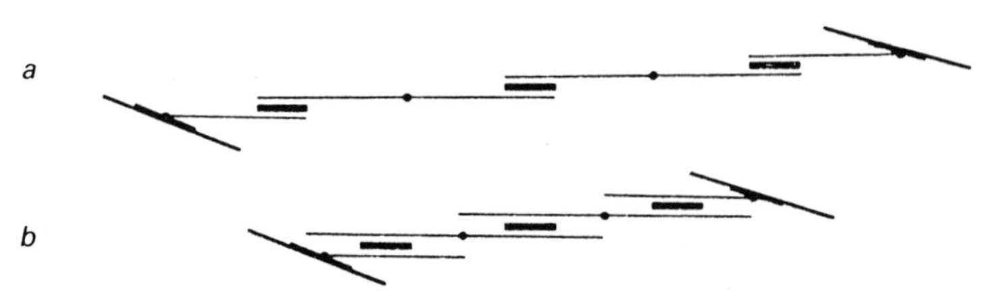

Fig. 11.8. A contractile unit of smooth muscle, showing how the filaments could slide past each other during contraction. From Squire (1986).

References and suggestions for further reading

Adrian, E. D. & Lucas, K. (1912). On the summation of propagated disturbances in nerve and muscle *J. Physiol., Lond.* **44**, 68–124.

Aidley, D. J. (1990). *The Physiology of Excitable Cells*, 3rd edn. Cambridge: University Press.

Armstrong, C. M. & Bezanilla, F. M. (1973). Current related to the movement of the gating particle of the sodium channels. *Nature, Lond.* **242**, 459–61.

Ashley, C. C. & Ridgeway, E. B. (1968). Simultaneous recording of membrane potential, calcium transient and tension in single muscle fibres. *nature, London.* **219**, 1168–9.

Baker, P. F., Hodgkin, A. L. & Shaw, T. I. (1962). The effect of changes in internal ionic concentrations on the electrical properties of perfused giant axons. *J. Physiol., London.* **164**, 355–74.

Barnard, E. A., Miledi, R. & Sumikawa, K. (1982). Translation of exogenous messenger RNA coding for nicotinic acetylcholine receptors produces functional receptors in *Xenopus* oocytes. *Proc. R. Soc. Lond.* **B215,** 241–6.

Bezanilla, F. (1985). Gating of sodium and potassium channels. *J. Membr. Biol.* **88**, 97–111.

Boyle, P. J. & Conway, E. J. (1941). Potassium accumulation in muscle and associated changes. *J. Physiol., Lond.* **100**, 1–63.

Brock, L. G., Coombs, J. S. & Eccles, J. C. (1952). The recording of potentials from motoneurones with an intracellular electrode. *J. Physiol., Lond.* **117**, 431–60.

Brown, M. C., Hopkins, W. G. & Keynes, R. J. (1991). *Essentials of Neural Development*. Cambridge: University Press.

Bülbring, E. (1979). Post junctional adrenergic mechanisms. *Brit. Med. Bull.* **35**, 285–94.

Buller, A. J. (1975). *The Contractile Behaviour of Mammalian Skeletal Muscle* (Oxford Biology Reader No. 36) London: Oxford University Press.

Burnstock, G. (1979). Autonomic innervation and transmission. *Brit. Med. Bull.* **35**, 255–62.

Cain, D. F., Infante, A. A. & Davies, R. E. (1962). Chemistry of muscle contraction. Adenosine triphosphate and phosphoryl creatine as energy supplies for single contractions of working muscle. *Nature, Lond.* **196**, 214–17.

Calwell, P. C. & Keynes, R. D. (1957). The utilization of phosphate bond energy for sodium extrusion from giant axons. *J. Physiol., Lond.* **137**, 12–13P.

Caldwell, P. C., Hodgkin, A. L., Keynes, R. D. & Shaw, T. I. (1960). The effects of injecting 'energy-rich' phosphate compounds on the active transport of ions in the giant axons of *Loligo*. *J. Physiol., Lond.* **152**, 561–90.

Catterall, W. A. (1986). Molecular properties of voltage-sensitive sodium channels. *Ann. Rev. Biochem.* **55**, 953–985.

Chandler, W. K. & Meves, H. (1970). Evidence for two types of sodium conductance in axons perfused with sodium fluoride solution. *J. Physiol., Lond.* **211**, 653–78.

Cole, K. S. & Curtis, H. J. (1939). Electric impedance of the squid giant axon during activity. *J. gen. Physiol.* **22**, 649–70.

Colquhoun, D. & Sakmann, B. (1985). Fast events in single-channel currents activated by acetylcholine and its analogues at the frog muscle end-plate. *J. Physiol., Lond.* **369**, 501–57.

Coombs, J. S., Eccles, J. C. & Fatt, P. (1955*a*). Excitatory synaptic action in motoneurones. *J. Physiol., Lond.* **130**, 374–95.

Coombs, J. S., Eccles, J. C. & Fatt, P. (1955*b*). The specific ionic conductances and the ionic movements across the motoneuronal membrane that produce the inhibitory postsynaptic potential. *J. Physiol., Lond.* **130**, 326–73.

Conway, E. J. (1957). Nature and significance of concentration relations of potassium and sodium ions in skeletal muscle. *Physiol. Rev.* **37**, 84–132.

Dale, H. H., Feldburg, W. & Vogt, M. (1936). Release of acetylcholine at voluntary motor nerve endings. *J. Physiol., Lond.* **86**, 353–80.

Davson, H. & Danielli, J. F. (1943). *The Permeability of Natural Membranes*. Cambridge: University Press.

Del Castillo, J. & Katz, B. (1954). Quantal components of the end-plate potential. *J. Physiol., Lond.* **124**, 560–73.

Del Castillo, J. & Katz, B. (1955). On the localization of acetylcholine receptors. *J. Physiol., London* **128**, 157–81.

Del Castillo, J. & Moore, J. W. (1959). On increasing the velocity of a nerve impulse. *J. Physiol., Lond.* **148**, 665–70.

DiFrancesco, D. & Noble, D. (1985). A model of cardiac electrical activity incoporating ionic pumps and concentration changes. *Phil. Trans R. Soc. Lond.* B **307**, 353–98.

Eccles, J. C. (1964). *The Physiology of Synapses*. Berlin: Springer Verlag.

Einthoven, W. (1924). The string galvanometer and measurement of the action currents of the heart. Nobel Lecture. Republished in 1965 in *Nobel lectures, Physiology or Medicine* 1921–41. Amsterdam: Elsevier.

Erlanger, J. & Gasser, H. S. (1937). *Electrical Signs of Nervous Activity*. Philadelphia: University of Pennsylvania Press.

Fatt, P. & Katz, B. (1951). An analysis of the end-plate potential recorded with an intracellular electrode. *J. Physiol., Lond.* **115**, 320–69.

Fatt, P. & Katz, B. (1952). Spontaneous subthreshold activity at motor nerve endings. *J. Physiol., Lond.* **117**, 109–28.

Frankenhaeuser, B. & Hodgkin, A. L. (1957). The action of calcium on the electrical properties of squid axons. *J. Physiol., Lond.* **137**, 218–44.
Glynn, I. M. (1988). How does the sodium pump pump? Chapter 1 in *Cell Physiology of Blood,* ed. R. C. Gunn & J. C. Parker. New York: Rockefeller University Press.
Gordon, A. M., Huxley, A. F. & Julian, F. J. (1966). The variation in isometric tension with sarcomere length in vertebrate muscle fibres. *J. Physiol., London.* **184**, 170–92.
Guy, H. R. (1988). A model relating the structure of the sodium channel to its functions. *Curr. Topics Membr. Transp.* **33**, 289–308.
Hamill, O. P., Marty, A., Neher, E., Sakmann, B. & Sigworth, F. J. (1981). Improved patch-clamp techniques for high-resolution current recording from cells and cell-free membrane patches. *Pflügers Arch.* **391**, 85–100.
Hill, A. V. (1938). The heat of shortening and the dynamic constants of muscle. *Proc. R. Soc. Lond.* **B126**, 136–95.
Hill, A. V. (1950*a*). A challenge to biochemists. *Biochim. biophys. Acta* **4**, 4–11.
Hill, A. V. (1950*b*). The dimensions of animals and their muscular dynamics. *Sci. Prog. Lond.* **38**, 209–30.
Hill, A. V. & Hartree, W. (1920). The four phases of heat production of muscle. *J. Physiol, Lond.* **54**, 84–128.
Hille, B. (1971). The hydration of sodium ions crossing the nerve membrane. *Proc. nat. Acad. Sci. U.S.A.* **68**, 280–2.
Hille, B. (1984). *Ionic Channels of Excitable Membranes.* Sunderland, massachusetts: Sinauer Associates.
Hille, B. (1989). Ionic channels: evolutionary origins and modern roles. *Q. J. exp. Physiol.* **74**, 785–804.
Hodgkin, A. L. (1939). The relation between conduction velocity and the electrical resistance outside a nerve fibre. *J. Physiol., Lond.* **94**, 560–70.
Hodgkin, A. L. (1951). The ionic basis of electrical activity in nerve and muscle. *Biol. Rev.* **26**, 339–409.
Hodgkin, A. L. (1958). Ionic movements and electrical activity in giant nerve fibres. *Proc. R. Soc. Lond.* **B148**, 1–37.
Hodgkin, A. L. (1975). The optimum density of sodium channels in an unmyelinated nerve. *Phil. Trans. R. Soc. Lond.* **B270**, 297–300.
Hodgkin, A. L. & Horowicz, P. (1957). The differential action of hypertonic solutions on the twitch and action potential of a muscle fibre. *J. Physiol., Lond.* **136**, 17–18P.
Hodgkin, A. L. & Horowicz, P. (1959). The influence of potassium and chloride ions on the membrane potential of single muscle fibres. *J. Physiol., Lond.* **148**, 127–60.
Hodgkin, A. L. & Horowicz, P. (1960). Potassium contractures in single muscle fibres. *J. Physiol, Lond.* **153**, 386–403.
Hodgkin, A. L. & Huxley, A. F. (1952). A quantitative description of membrane current and its application to conduction and excitation in nerve. *J. Physiol, Lond.* **117**, 500–44.

Hodgkin, A. L., Huxley, A. F. & Katz, B. (1952). Measurement of current voltage relations in the membrane of the giant axon of *Loligo*. *J. Physiol., Lond.* **116**, 424–48.

Hodgkin, A. L. & Katz, B. (1949). The effects of sodium ions on the electrical activity of the giant axon of the squid. *J. Physiol., Lond.* **108**, 37–77.

Hodgkin, A. L. & Keynes, R. D. (1955). Active transport of cations in giant axons from *Sepia* and *Loligo*. *J. Physiol., Lond.* **128**, 28–60.

Homsher, E. (1987). Muscle enthalpy production and its relationship to actomyosin ATPase. *Ann. Rev. Physiol.* **49**, 673–90.

Huxley, A. F. (1980). *Reflections on Muscle*. Liverpool: University Press.

Huxley, A. F. & Niedergerke, R. (1954). Structural changes in muscle during contraction. Interference microscopy of living muscle fibres. *Nature, Lond.* **173**, 971–3.

Huxley, A. F. & Stämpfli, R. (1949). Evidence for saltatory conduction in peripheral myelinated nerve fibres. *J. Physiol., Lond.* **108**, 315–39.

Huxley, A. F. & Taylor, R. E. (1958). Local activation of striated muscle fibres. *J. Physiol., Lond.* **144**, 426–41.

Huxley, H. E. (1963). Electron microscope studies on the structure of natural and synthetic protein filaments from striated muscle. *J. mol. Biol.* **7**, 281–308.

Huxley, H. E. (1971). The structural basis of muscular contraction. *Proc. R. Soc. Lond.* B**178**, 131–49.

Huxley, H. E. (1976). The structural basis of contraction and regulation in skeletal muscle. In *Molecular Basis of Motility*, ed. L. M. G. Heilmeyer Jr, J. C. Ruegg & Th. Wieland. Berlin: Springer-Verlag.

Huxley, H. E. & Hanson, J. (1954). Change in the cross-striations of muscle during contraction and stretch and their structural interpretation. *Nature, Lond.* **173**, 973–6.

Infante, A. A. & Davies, R. E. (1962). Adenosine triphosphate breakdown during a single isotonic twitch of frog sartorious muscle. *Biochem. biophys. Res. Commun.* **9**, 410–15.

Katz, B. & Miledi, R. (1965). The measurement of synaptic delay, and the time course of acetylcholine release at the neuromuscular junction. *Proc. R. Soc. Lond.* B**161**, 483–95.

Katz, B. & Miledi, R. (1967). The timing of calcium action during neuromuscular transmission. *J. Physiol., Lond.* **189**, 535–44.

Keynes, R. D. (1951). The ionic movements during nervous activity. *J. Physiol., Lond.* **114**, 119–50.

Keynes, R. D. (1963). Chloride in the squid giant axon. *J. Physiol., Lond.* **169**, 690–705.

Keynes, R. D. (1983). Voltage-gated ion channels in the nerve membrane. *Proc. R. Soc. Lond.* B**220**, 1–30.

Keynes, R. D. (1989). The role of giant axons in studies of the nerve impulse. *BioEssays* **10**, 90–3.

Keynes, R. D. (1990). A series-parallel model of the voltage-gated sodium channel. *Proc. R. Soc. Lond.* B**240**, 425–32.

Keynes, R. D. (1991). On the voltage dependence of inactivation the sodium channel of the squid giant axon. *Proc. R. Soc. Lond.* **B243**, 47–53.

Keynes, R. D., Greeff, N. G. & Forster, I. C. (1990). Kinetic analysis of the sodium gating current in the squid giant axon. *Proc. R. Soc. Lond.* **B240**, 411–23.

Keynes, R. D. & Lewis, P. R. (1951). The sodium and potassium content of cephalopod nerve fibres. *J. Physiol., Lond.* **114**, 151–82.

Keynes, R. D. & Martins-Ferreira, H. (1953). Membrane potentials in the electroplates of the electric eel. *J. Physiol., Lond.* **119**, 315–51.

Keynes, R. D. & Ritchie, J. M. (1984). On the binding of labelled saxitoxin to the squid giant axon. *Proc. R. Soc. Lond.* **B222**, 147–53.

Keynes, R. D. & Rojas, E. (1974). Kinetics and steady-state properties of the charged system controlling sodium conduction in the squid giant axon. *J. Physiol., Lond.* **239**, 393–434.

Kubo, T., Fukuda, K., Mikami, A., Maeda, A., Takahashi, H., Mishina, M., Haga, T., Haga, K., Ichiyama, A., Kangawa, K., Kojima, M., Matsuo, H., Hirose, T. & Numa, S. (1986). Cloning, sequencing and expression of complementary DNA encoding the muscarinic acetylcholine receptor. *Nature, Lond.* **323**, 411–16.

Kuffler, S. W. (1980). Slow synaptic responses in the autonomic ganglia and the pursuit of a peptidergic transmitter. *J. exp. Biol.* **89**, 257–86.

Kyte, J. & Doolittle, R. F. (1982). A simple method for displaying the hydropathic character of a protein. *J. mol. Biol.* **157**, 105–32.

Loewi, O. (1921). Über humorale Übertragbarkeit der Herznervenwirkung. *Pflügers Arch. ges. Physiol.* **189**, 239–42.

Makowski, L., Caspar, D. L. D., Phillips, W. C., Baker, T. S. & Goodenough, D. A. (1984). Gap junction structures VI. Variation and conservation in connexon conformation and packing. *Biophys. J.* **45**, 208–18.

Merton, P. A. (1954). Voluntary strength and fatigue. *J. Physiol, Lond.* **128**, 553–64.

Merton, P. A., Hill, D. K. & Morton, H. B. (1981). Indirect and direct stimulation of fatigued human muscle. In *Human Muscle Fatigue: Physiological Mechanisms*, ed. R. Porter & J. Whelan, pp. 120–6. London: Pitman Medical.

Neher, E. & Sakmann, B. (1976). Single-channel currents recorded from membrane of denervated frog muscle cells. *Nature, Lond.* **260**, 799–802.

Noble, D. (1979). *The Initiation of the Heartbeat*, 2nd edn. Oxford: University Press.

Noble, D. (1984). The surprising heart: a review of recent progress in cardiac electrophysiology. *J. Physiol., Lond.* **353**, 1–50.

Noda, M., Ikeda, T., Kayano, T., Suzuki, H., Takeshima, H., Kurasaki, M., Takahashi, H. & Numa, S. (1986). Existence of distinct sodium channel messenger RNAs in rat brain. *Nature, Lond.* **320**, 188–92.

Noda, M., Shimizu, S., Tanabe, T., Takai, T., Kayano, T., Ikeda, T., Takahashi, H., Nakayama, H., Kanaoka, Y., Minamino, N., Kangawa, K., Matsuo, H., Fatery, M. A., Hirose, T., Inayama, S., Hayashida, H.,

Miyati, T. & Numa, S. (1984). Primary structure of *Electrophorus electricus* sodium channel deduced from cDNA sequence. *Nature, Lond.* **312**, 121–7.

Noda, M., Takahashi, H., Tanabe, T., Toyosato, M., Furutani, Y., Hirose, T., Asai, M., Inayama, S., Miyata, T. & Numa, S. (1982). Primary structure of α-subunit precursor of *Torpedo californica* acetylcholine receptor deduced from cDNA sequence. *Nature, Lond.* **279**, 793–7.

Offer, G. (1974). The molecular basis of muscular contraction. In *Companion to Biochemistry*, ed. A. T. Bull, J. R. Lagnado, J. O. Thomas and K. F. Tipton, pp. 623–71. London: Longman.

Olsen, R. W. & Tobin, A. J. (1990). Molecular biology of GABA-A receptors. *FASEB J.* **4**, 1469–80.

Peachey, L. D. (1965). The sarcoplasmic reticulum and transverse tubules of the frog's sartorius. *J. Cell Biol.* **25**, 209–32.

Ritchie, J. M. & Rogart, R. B. (1977). The binding of saxitoxin and tetrodotoxin to excitable tissue. *Rev. Physiol. Biochem. Pharmacol.* **79**, 1–50.

Robertson, J. D. (1960). The molecular structure and contact relationships of cell membranes. *Prog. Biophys.* **10**, 343–418.

Rushton, W. A. H. (1933). Lapicque's theory of curarization. *J. Physiol., Lond.* **77**, 337–64.

Ryall, R. W. (1979). *Mechanisms of Drug Action on the Nervous System.* Cambridge: University Press.

Sakmann, B. & Neher, E. (1983). *Single-Channel Recording.* Plenum Press: New York.

Salkoff, L., Butler, A., Wei, A., Scavarda, N., Baker, K., Pauron, D. & Smith, C. (1987). Molecular biology of the voltage-gated sodium channel. *Trends Neurosci.* **10**, 522–7.

Scher, A. M. (1965). Mechanical events in the cardiac cycle. In *Physiology and Biophysics*, ed. T. C. Ruch & H. D. Patten. Philadelphia: Saunders.

Scher, A. M. (1974). Electrocardiograms. In *Physiology and Biophysics*, 20th edn: *Circulation, Respiration and Fluid Balance*, ed. T. C. Ruch, H. D. Patten & A. M. Scher. Philadelphia: W. B. Saunders.

Schmidt-Nielsen, K. (1990). *Animal Physiology*, 4th edition. Cambridge: University Press.

Schofield, P. R., Darlison, M. G., Fujita, M., Burt, D. R., Stephenson, F. A., Rodriguez, H., Rhee, L. M., Ramachandran, J., Reale, V., Glencorse, T. A., Seeburg, P. H. & Barnard, E. A. (1987). Sequence and functional expression of the $GABA_A$ receptor shows a ligand-gated receptor super-family. *Nature, Lond.* **328**, 221–7.

Skou, J. C. (1957). The influence of some cations on an adenosine triphosphatase from peripheral nerves. *Biochim. Biophys. Acta* **23**, 394–401.

Squire, J. M. (1986). *Muscle: Design, Diversity and Disease.* Menlo Park, California: Benjamin/Cummings.

Takeuchi, A. & Takeuchi, N. (1959). Active phase of frog's end-plate potential. *J. Neurophysiol.* **22**, 395–411.

Takeuchi, A. & Takeuchi, N. (1960). On the permeability of the end-plate membrane during the action of the transmitter. *J. Physiol., Lond.* **154**, 52–67.

Tasaki, I. (1953). *Nervous Transmission*. Springfield.

Weidmann, S. (1956). *Elektrophysiologie der Herzmuskelfaser*. Huber: Berne.

Whittaker, V. P., Dowdall, M. J. & Boyne, A. F. (1972). The storage and release of acetylcholine by cholinergic nerve terminals: recent results with non-mammalian preparations. *Biochem. Soc. Symp.* **36**, 49–68.

Wilkie, D. R. (1968). Heat work and phosphorylcreatine breakdown in muscle. *J. Physiol., Lond.* **195**, 157–83.

Wilkie, D. R. (1976). Energy transformation in muscle. In *Molecular Basis of Motility*, ed. L. M. G. Heilmeyer Jr., J. C. Ruegg & Th. Wieland, pp. 69–80. Berlin: Springer-Verlag.

Zagotta, W. N. & Aldrich, R. W. (1990). Voltage-dependent gating of *Shaker* A-type potassium channels in *Drosophila* muscle. *J. gen. Physiol.* **95**, 29–60.

Index